INFECTION WITH NON-SPORING ANAEROBIC BACTERIA

INFECTION WITH NON-SPORING ANAEROBIC BACTERIA

A Symposium of the British Society for Antimicrobial Chemotherapy held at the Royal College of Physicians, London

Edited by
IAN PHILLIPS M.A., M.D., M.R.C. Path.
Professor of Medical Microbiology,
St. Thomas's Hospital Medical School, London
and
MAX SUSSMAN B.Sc., Ph.D., M.I.Biol.
Senior Lecturer in Medical Microbiology,
Welsh National School of Medicine, Cardiff

CHURCHILL LIVINGSTONE
EDINBURGH LONDON AND NEW YORK
1974

CHURCHILL LIVINGSTONE

Medical Division of Longman Group Limited
Distributed in the United States of America by Longman Inc.,
New York and by associated companies, branches and
representatives throughout the world.

First published 1974
ISBN 0 443 01243 1

Printed in Great Britain

PREFACE

There is an increasing recognition of the importance of the non-sporing anaerobic bacteria in human infections. The object of this symposium was to review the present state of knowledge about these organisms and the part they play in human disease. It was organised in order to bring together bacteriologists and clinicians to discuss common problems, and the accent is therefore on diagnosis and treatment.

We are particularly grateful to Mrs Dawn Harris for her help in organising the Symposium and to Upjohn Limited, Crawley, Essex, for financial support.

IAN PHILLIPS,
Department of Microbiology.
St Thomas's Hospital Medical School,
London, SE1 7EH.

MAX SUSSMAN,
Department of Medical Microbiology,
Welsh National School of Medicine,
Cardiff.

1974

LIST OF PARTICIPANTS

Symposium on Infection with Non-sporing Anaerobic Bacteria

February, 1974

Dr J.G. Barlett
Veterans Administration Hospital, Sepulveda, California, and Department of Medicine, UCLA School of Medicine, Los Angeles, California.

*Dr P.A. Boswell
King's College Hospital Medical School, Denmark Hill, London, SE5 8RX.

*Professor W. Brumfitt
Department of Medical Microbiology, Royal Free Hospital School of Medicine, Gray's Inn Road, London WC1X 8LF.

Dr J.G. Collee
Department of Bacteriology, University of Edinburgh, Teviot Place, Edinburgh EH8 9AG.

*Dr B.I. Davies
Edgeware General Hospital, Pathology Department, Edgeware, Middlesex.

*Dr E.G. Dowsett
Basildon Hospital, Nethermayne, Basildon, Essex.

Dr B. Drasar
Bacterial Metabolism Research Lab., Public Health Laboratory Service, Colindale Avenue, London NW9 5DX.

Dr I.R. Ferguson
Public Health Laboratory, Luton and Dunstable Hospital, Luton, LU4 0DZ.

Dr S.M. Finegold
Chief, Infectious Disease Section, Wadsworth Hospital Center, Wilshire and Sawtelle Boulevards, Los Angeles, California 90073.

*Professor L.P. Garrod
Stradbroke, Gipsy Lane, Wokingham, Berks RG11 2HP.

*Dr A.T. Ghoneim
Public Health Laboratory, Church Lane, Heavitree, Exeter.

*Dr S. Goodwin
Clinical Research Centre, Northwick Park Hospital, Watford Road, Harrow, Middlesex HA1 3UJ.

Mr J. Hardie
Dental Bacteriology Laboratory and MRC Dental Epidemiology Unit, The London Hospital Medical College, London E1 2AD.

Dr J. Harrison
Frenchay Hospital, Bristol.

*Dr R. Jacomb
Upjohn Limited, Fleming Way, Crawley, Sussex.

*Professor H. Lambert
St George's Hospital, Blackshaw Road, London SW17.

*Dr A.C. Maddocks
St Mary's Hospital Medical School, Department of Bacteriology, Paddington, London W2 1PG.

*Dr L.S. Nakhla
Department of Clinical Pathology, St Albans City Hospital, Normandy Road, St Albans, Herts.

*Dr S.W.B. Newson
Papworth Hospital, Papworth Everard, Near Cambridge, CB3 8RE.

*Professor F. O'Grady
Department of Bacteriology, St Bartholomew's Hospital, London, EC1.

*Dr O.A. Okubadejo
Public Health Laboratory, St Mary's Hospital, East Wing, Milton Road, Portsmouth PO3 6AQ.

Dr A. Percival
Department of Medical Microbiology, The University of Liverpool. New Medical School, Pembroke Place, Liverpool L69 3BX.

Professor I. Phillips
Department of Microbiology, St Thomas's Hospital Medical School, London SE1 7EH.

*Mr A.A. Quayle
11 Wentwood Drive, Llanvaches, Newport, Monmouthshire.

*Dr D.S. Reeves
Microbiology Department, Southmead Hospital, Bristol.

*Dr P.J. Say
School of Pathology, Middlesex Hospital Medical School, Riding House Street, London W1P 7LD.

*Dr D.C. Shanson
Department of Microbiology, University College Hospital, Grafton Way, London WC1 6AU.

*Dr S. Selwyn
Department of Bacteriology, Westminster Hospital Medical School, London S.W.1.

*Dr J.D. Sleigh
Western Infirmary, Glasgow.

*Dr R.C. Spencer
Bacteriology Department, Wright Fleming Institute, St Mary's Hospital, London.

*Dr E.J. Stokes
Department of Microbiology, University College Hospital, Grafton Way, London WC1 6AU.

Dr S. Tabaqchali
Department of Medical Microbiology, St Bartholomew's Hospital, West Smithfield, London EC1A 7BE.

Dr B. Watt
Central Microbiological Laboratories, Western General Hospital, Crewe Road, Edinburgh EH4 2XU.

Dr R.E.O. Williams
Director, Public Health Laboratory Service, Colindale Avenue, London NW9 5HT.

Dr A.T. Willis
Public Health Laboratory, Luton and Dunstable Hospital, Lewsy Road, Luton LU4 ODZ

*Dr R.L. Willson
Biochemistry Department, Brunel University, Uxbridge, Middlesex.

Dr S. Young
Public Health Laboratory, Luton and Dunstable Hospital, Luton LU4 ODZ

* Contributors to Discussion and Chairman

CONTENTS

1. Introduction

R. E. O. WILLIAMS

We are here today to talk about infections with a group of non-sporing bacteria that will not grow under aerobic conditions. The recognition of such pathogenic microbes goes back a long way and in fact *Bacillus necrophorus* was described from calf diphtheria in 1884 only one year after Klebs had found his bacillus in human diphtheria; by 1896 Vincent had described his famous group of organisms in the mouth and about that time the anaerobes of the gut had started to acquire the enormous variety of names that they have suffered from ever since. The last edition of Topley & Wilson, published 10 years ago, says that classification was difficult and justifies that very amply by the next few pages describing the various systems proposed between 1884 and 1964, and at that time the additions from Virginia had not in fact reached the editors. But let us hope that one of the things that may come out from the discussions today may be some clarification of some of these nomenclatural points.

I suppose that 10 years ago most medical microbiologists would have interpreted anaerobic infections as infections due to clostridia, just as in the same way most doctors and a good many medical microbiologists would have regarded the coliform group of organisms as the dominant bacteria of the gut. But clearly we would not be here today if it were not that all this has changed and that non-sporing anaerobes are now 'in'. As a matter of fact, what appears to be 'in' is the group of Gram-negative non-sporing rods, typically *Bacteroides*, and I think it is worth remembering that these are only one of the groups of non-sporing anaerobic bacteria which range from *Actinomyces* at one end to a whole lot of ill-defined anaerobic cocci, at the other. I hope that during the course of the day we shall be hearing something about the possible relevance or irrelevance of some of the anaerobic, non-sporing bacteria, other than *Bacteroides* itself.

Clearly one of the objectives of the day must be to review the changes in microbiological techniques that have permitted and encouraged this change in outlook. The other must be to discuss how this greater interest and understanding of the bacteria themselves helps us to understand and manage disease in man. Certainly one of the reasons why we are only recently coming to any great degree of interest in non-sporing anaerobic bacteria has been their relative reluctance to survive in the sort of specimens we are used to receiving from our patients in the wards, or to grow on the sort of culture media and under the sort of cultural conditions that proved so profitable for isolating the good old-fashioned pathogens like staphylococci, streptococci, coliforms and so forth. Like so many techniques the ability to grow the anaerobic organisms is not so difficult when you know how, and we shall doubtless hear during the course of this morning some useful tricks from Drs Watt, Collee and Draser who have been vying with each other over the last few years, each describing his method and hoping to show that it is either better, quicker, cheaper, or more expensive or more elaborate or just more enjoyable, than the other's. The consequences of being able to isolate these organisms is that you do in fact isolate them from a very large number of situations. We now know that the Gram-negative non-sporing rods are present in the intestinal canal of all healthy adult men at counts of something to the order of 10^{10} or 10^{11} per gram, and forming some 10 per cent of the wet weight of faeces.

Probably, during the day, we shall hear a good many people say that these organisms were'the cause' of some particular condition and I hope that mentally (physically would be better, but mentally is all that we are likely to achieve) on these occasions a bell will ring which will ask you to say, 'do you mean was the cause of or do you mean was isolated from'? We do have a real problem of considering the difference between mere presence of the bacteria and relevance to disease. One reason why this is complicated is that the kind of infections that these organisms can produce, and the kind of disease that they can cause, are not always straightforward. Sometimes you do find anaerobic Gram-negative rods in pure culture in pyogenic lesions in a situation where you can say that they are the cause of the infection. But much more often they are present in mixed culture, with or without aerobes.

The veterinary microbiologists were in the field of the anaerobic non-sporing bacteria in advance of the medical microbiologists, studying the bovine rumen and its bacterial flora, and were very soon unravelling metabolic interactions between the aerobes and the anaerobes and between the anaerobes themselves. There is no doubt that many of the *Bacteroides* and similar organisms are metabolically extremely active and very versatile, and there is growing evidence that in the human gut, as in the animal gut, these organisms play an important part in maintaining the normal physiological state by recycling material secreted into the gut in one way or another. The metabolic interactions have also been demonstrated in a number of infective lesions. Some time ago Macdonald made a classical demonstration of the activity of *Bacteroides melaninogenicus* as a pathogen, but only in circumstances where it was provided with appropriate growth factors and growth opportunities by the activities of other bacteria; and Socransky and Gibbons demonstrated pathogenic activities by various oral bacteria that only took place in the presence of *Bacteroides*. So the big problem that I see for the afternoon session will be to define the circumstances in which the isolation of anaerobes from disease processes implies that they cause the disease. It will be interesting to see whether we can, in fact, come across any situations where the kind of bacterial synergy that I have mentioned means that in fact our treatment should not necessarily be directed at the proximate pathogen, but directed perhaps at another organism that may be producing the conditions in which the pathogen can work. It will be an additional activity for chemotherapy if we do not have to treat the microbe that causes the disease but only the one that causes the condition in which some other microbe can cause disease.

LABORATORY ASPECTS OF ANAEROBIC INFECTIONS

2. Practical Approaches to the Isolation and Identification of Clinically Important Non-sporing Anaerobes

B. WATT and J. G. COLLEE

THE ISOLATION OF NON-SPORING ANAEROBES

The non-sporing anaerobes, notably the Gram-negative anaerobic bacilli (bacteroides-like organisms) and the anaerobic cocci, are becoming increasingly recognised as potential pathogens associated with infection in hospital patients. The isolation and identification of these organisms from clinical material often poses problems for the clinical microbiologist. The present paper seeks to suggest practical approaches to the recovery of these organisms in the laboratory, from the point of view of the diagnostic microbiologist who is attempting to give his clinical colleagues timely and useful bacteriological information relevant to the management of the patient.

Most bacteriologists attempting to recover anaerobes from clinical material would agree that, whereas growth is relatively easily obtained in a suitable liquid medium, it may be much more difficult to achieve prompt and reliable growth on solid media. We accordingly visualise a 'scale of despair' that takes account of the current ease or difficulty involved in the laboratory culture of anaerobic bacteria on solid media (Collee and Watt, 1972). Such a scale might show *Clostridium histolyticum* (an 'easy' organism) at one extreme and *Clostridium oedematiens* type D, one of the more exacting anaerobes, at the other. The bacteroides-like organisms and the anaerobic cocci seem to occupy an intermediate position, as indicated in Fig. 2.1. However, the position that we allocate to any organism on this scale merely reflects our present knowledge and laboratory experience. For example, our work has shown that different organisms behave differently with different reducing agents (Watt, 1972*a*); perhaps we cannot hope to find a reagent that uniformly improves the growth of all anaerobes and thus moves them *en bloc* up the scale.

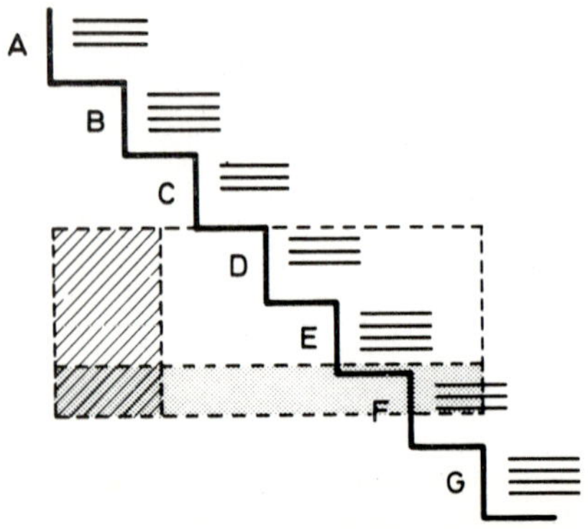

Fig.2.1. A concept of graded strictness in terms of requirement for anaerobic or reducing conditions for growth; the figure represents a descending scale of difficulty experienced in the laboratory culture of different (un-named) groups of anaerobic bacteria on solid media. Organisms of the *Bacteroides* group might be allocated to the area shown by the broken line.

In the case of the bacteroides-like organisms, which until quite recently were regarded as 'difficult', recovery can be made generally reliable and certainly more prompt, if account is taken of the following points.

ROUTINE LABORATORY PROCEDURES

A defined anaerobic procedure

We have found that a standard anaerobic procedure, as outlined in Fig.2.2, gives good results both in a busy diagnostic laboratory and for research purposes. We use the anaerobic jar marketed by Baird and Tatlock Ltd (BTL) and we now fit it with three catalyst sachets as a routine; the jar is evacuated, filled with a mixture of 90 per cent H_2 and 10 per cent CO_2, and left on the bench for 10 min. at room temperature. It is essential to follow all of the steps shown in Fig.2.2 and Appendix B with strict attention to detail. The prompt demonstration of a secondary vacuum of at least 20 mm Hg in a processed jar gives a reliable index of catalyst activity. Defective catalysts or leaks in the jars are revealed by the manometry step, thus allowing early diagnosis of faults and avoiding loss of an important culture. The disadvantage of various types of indicators of the anaerobic state is that they are slow to reveal faults; the failure of a jar may not be evident until a period ranging from a few hours to several days has elapsed. Moreover, as the rubber connection between the indicator tube and the side-arms on BTL jars commonly develop a leak, we dispense with the external indicator, remove the rubber and seal the side-arm. The provision of three catalyst sachets per jar has been shown to improve the recovery of some anaerobes and to reduce 'jar variation' (Watt, Hoare and Collee, 1973).

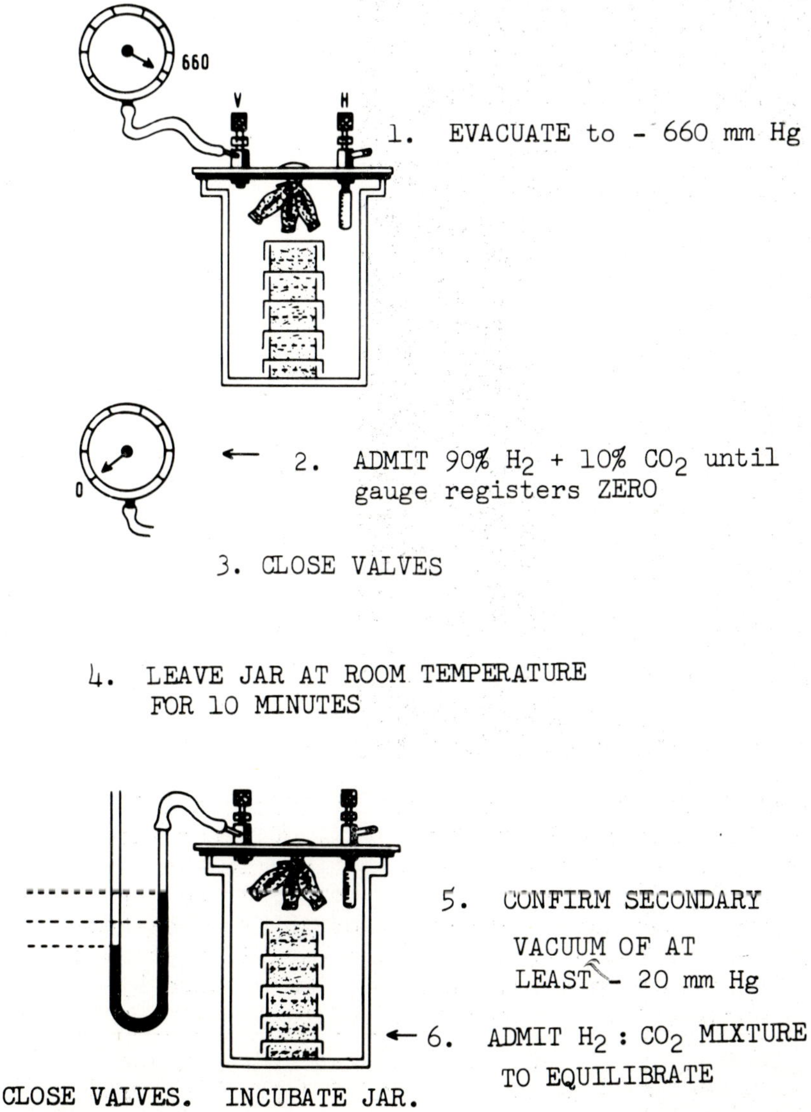

Fig.2.2. A defined procedure for use with the Baird and Tatlock (BTL) anaerobic jar and a cylinder source of hydrogen and carbon dioxide. The plastic petri dishes should be of the vented type. We recommend that three catalyst sachets should be fitted to each jar as a routine. The manometry step provides a prompt indication that the catalyst is working.

A vacuum pump and cylinders of compressed gases are required for our standard anaerobic procedure. For laboratories in which experience of anaerobic bacteriology is limited, the Gaspak System (Fig. 2.3) offers a practical alternative, without the need for gas cylinders and pumps.

Fig. 2.3. The Gaspak disposable source of hydrogen and carbon dioxide is activated by putting 10 ml of water into the foil pack. Some of the studies summarised in the present paper employed this system with the Baird and Tatlock (BTL) anaerobic jar shown; most of the studies were done with the BTL jar and a standardised anaerobic procedure (p. 9). Note that an external indicator is not used by the Edinburgh workers; the side-arm is sealed.

The Gaspak disposable foil envelope produces hydrogen and carbon dioxide when 10 ml of water is added. We have used it in conjunction with the BTL jar, and have found it to compare well with our standard procedure in quantitative studies (Collee *et al.*, 1972). The results of a quantitative experiment with non-sporing anaerobes are given in Table 2.1 and it can be seen that both systems gave good growth of the *Bacteroides* test strains.

Table 2.1. A comparison of results obtained in quantitative studies with a standardised anaerobic procedure and with the Gaspak system. In both series, Baird and Tatlock anaerobic jars were used.

Test organism	Inoculum*	Estimated viable count per ml derived from three replicate plates incubated in	
		A	B
Bacteroides fragilis NCTC 9343	0.02 ml of a dilution of 1 in 1 million	3.3×10^9	3.8×10^9
B. necrophorus NCTC 10575	0.02 ml of a dilution of 1 in 10,000	1.1×10^7	1.6×10^7

*20-hr cooked meat broth cultures were used.
A = Standard anaerobic procedure using BTL jar;
B = Gaspak system used in conjunction with BTL jar.

The provision of carbon dioxide

Both the standard anaerobic procedure and the Gaspak ensure that carbon dioxide is provided in the anaerobic atmosphere, and this is important. Qualitative studies with anaerobes have shown that CO_2 at a concentration of 10 per cent enhances the speed of recovery and size of colonies grown on solid media. The results of our quantitative studies, illustrated in Table 2.2, confirm that when CO_2 is used, there is often a marked improvement, and sometimes a dramatic improvement, in the numbers of organisms recovered.

Table 2.2. Influence of carbon dioxide on recovery of Bacteroides organisms from dilutions of cooked-meat broth cultures.

Test organism	Inocula	Mean colony count ± S.E. derived from ten replicate plates* incubated in A	B	Significance
18-hr CMB culture of *Bacteroides*† *fragilis* NCTC 9343	0.02 ml of the test cultures at a dilution of 1 in 2000	1153.0±32.4	873.0±26.4	$p<0.001$
18-hr CMB culture of *B. necrophorus* NCTC 7155		739.6±21.2	17.8±3.5	$p<0.001$

*10 per cent equine blood agar plates were used
A = 10 per cent. CO_2 + 90 per cent. H_2
B = 100 per cent. H_2
p derived from t-test for difference of means on 18 degrees of freedom
†The identity of this strain is debated at present.

Use of suitable culture media

The currently recognised anaerobic pathogens of man can be grown on blood agar media, and good quantitative recoveries can be achieved especially if the media are freshly prepared. The use of solid media that have been stored aerobically is not recommended for the recovery of exacting anaerobes (Watt, 1972*a*). An adequately nutritious base, such as Columbia blood agar base (Oxoid Ltd) is essential for the satisfactory growth of demanding anaerobes such as some *Bacteroides* strains and anaerobic cocci. Blood (10 per cent) should be added to the base; human blood gives marginally better results, but horse blood is also good (Collee and Watt, *unpublished results*). Specific supplements may enhance the growth of certain organisms; for example, we have found that the cysteine-dithiothreitol system of Moore (1968) can be usefully incorporated in a blood agar medium for the growth of exacting strains of *Cl. oedematiens*, but this supplement did not enhance the growth of other clinically important anaerobes that we tested (Watt, 1972*b*). Other workers have claimed that special supplements should be added as a routine for the growth of certain *Bacteroides* strains. We have not been able to demonstrate particular requirements for menadione or haemin that are not met by an adequately nutritious blood agar medium (but see Bowden and Hardie, 1971).

Selection of selective media

Selective media are widely employed for the isolation of anaerobes, and the wide range of inhibitory agents that have been suggested as key constituents of selective media is testimony to the fact that no 'ideal' selective medium has so far been devised. The subject has been reviewed by Finegold, Sugihara and Sutter (1971) (and see Chapter 4). The range of inhibitory agents includes neomycin, kanamycin, vancomycin, bile salts and their derivatives. In the hospital service laboratory, selective media should be used in conjunction with non-selective media as most if not all selective media will have some inhibitory effect against the selected species. We have found that neomycin blood agar is a useful selective medium for bacteroides-like organisms and anaerobic cocci, provided that the concentration of neomycin is kept to 70 μg per ml; higher concentrations inhibited recovery of non-sporing anaerobes from clinical material (Watt, 1972*a*).

DEBATED REQUIREMENT FOR SPECIALISED ANAEROBIC PROCEDURES

Several workers have suggested that specialised procedures such as the use of anaerobic cabinets or of oxygen-free gas flushing devices, are essential for the recovery of demanding anaerobes on solid media. As we were able to obtain good recoveries of exacting sporing and non-sporing anaerobes of clinical importance by our standardised procedure or by the use of the Gaspak with a BTL anaerobic jar at the bench, we have been reluctant to accept this view. The results of our subsequent studies with an anaerobic cabinet have strengthened our reservations. Full details of the design and construction of the cabinet are already in press (Watt, Collee and Brown, 1974). A modified steel glove box was flushed with a mixture of nitrogen and hydrogen, with CO_2 at a final concentration of 10 per cent. This gas mixture was rendered oxygen-free by passage through a catalytic purifier and the atmosphere of the cabinet was demonstrably anaerobic.

We compared the recoveries obtained with a series of test organisms from samples processed inside and outside the cabinet. For example, the data obtained in studies of the recoveries of total anaerobes (mainly bacteroides-like organisms) from faecal suspensions are shown in Table 2.3. The use of the cabinet did not improve the yields obtained.

Table 2.3. Recovery of total anaerobes from human faecal samples processed within an anaerobic cabinet compared with that from samples processed with a standard anaerobic procedure.

Test inocula	Mean surface viable counts ± S.E. derived from samples processed	
	In the anaerobic cabinet	By a standard anaerobic procedure at the bench
0.02 ml of samples diluted 1 in 10 million	$(2.4\pm0.4)^* \times 10^{10}$	$(2.4\pm0.3) \times 10^{10}$

*Counts expressed as total anaerobes per g.

PROBLEMS ASSOCIATED WITH THE COLLECTION AND SUBMISSION OF CLINICAL SPECIMENS

Good anaerobic procedure cannot redeem organisms that have been killed in transit, nor can it compensate for inadequate sampling. Too often, a swab is regarded as a convenient method of sampling from a clinical situation; fragments of tissue or a small volume of pus, transported promptly to the laboratory, are much more likely to yield a representative sample.

Material submitted on a swab is immediately exposed to the hazards of aerobic exposure and desiccation, and it is well known that such specimens are often considerably delayed in transit to the laboratory. Points in the chain of submission of clinical specimens to the laboratory at which loss or inactivation of anaerobic pathogens might occur are now being defined (see Fig.2.4). With the *Bacteroides* and *Clostridium* models that we have studied so far, it appears that considerable losses are attributable to the general use of swabs for sampling wound exudates. It has been assumed, on the basis of good evidence with aerobic pathogens, that the swab material may be toxic or that aerobic exposure is the principal toxic factor responsible for the inactivation of anaerobes in transit. Our present evidence (Collee *et al.*, 1974) suggests that desiccation may be of more importance and that remarkably inefficient release of organisms from the swab on to a plate of solid medium is a very significant factor in the failure of quantitative recovery of these organisms on primary culture.

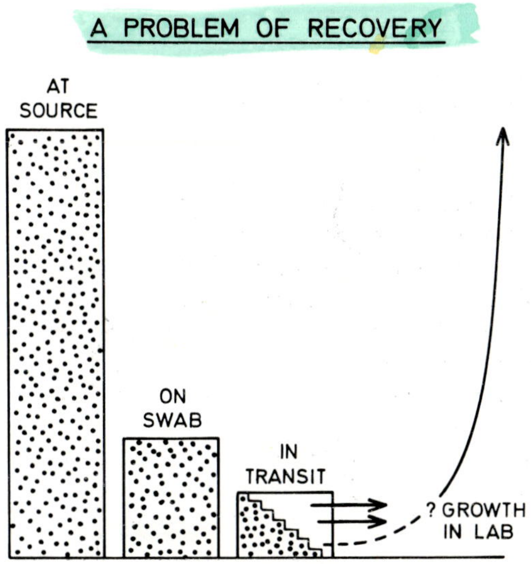

Fig.2.4. Sequential losses in transit: A proportion of the potential number of organisms present in a wound exudate is sampled on a swab. Various adverse factors then contribute to a decrease in the viable numbers on the swab, and relatively few of these viable cells are released from the swab to the primary plate. The problem of achieving prompt and reliable growth of the survivors in the laboratory is perhaps made unduly difficult by our present practices in submitting and handling specimens for clinical anaerobic bacteriology.

There is a need for the more general use of suitable transport media, provided that the submission of a specimen in this manner is not taken as an invitation for further delay in submission and processing. Our preliminary studies with thioglycollate seem to indicate that a transport medium containing this agent at a concentration of 1 to 1.5 per cent (see for example, Stuart, 1956; and Cary and Blair, 1964) would be acceptable.

THE CHARACTERISATION OF THE GRAM-NEGATIVE NON-SPORING ANAEROBIC BACILLI

There are various approaches to the detailed characterisation of this difficult group of organisms, and these will be dealt with in detail elsewhere in this book (see Appendix C). However, the clinical microbiologist is obliged to give a reasonably accurate report with reasonable speed. Since time does not permit unduly detailed identification procedures to be undertaken as a routine, we suggest that the following information is of value:

1. *Microscopic morphology* of the organism.
2. *Demonstration of requirement for strictly anaerobic conditions* for growth.
3. *Colonial appearance* on blood agar media; the presence or absence of haemolysis and pigment production.
4. *Biochemical properties* in tests with pure cultures including observations of: fermentative properties with glucose, lactose, sucrose and maltose; production of hydrogen sulphide; formation of indole; hydrolysis of aesculin. There are, however, technical difficulties involved in obtaining reproducible results in these tests (see Collee, 1974).
5. *Antibiotic sensitivity and resistance patterns*. These can provide clues to identification, in addition to their practical clinical value. The stability of the antibiograms for particular species in this group merits further investigation and there are problems in relating disk sensitivity test results to estimates of minimum inhibitory concentrations of some antibiotics (see Chapter 4).
6. *Special tests*. Determination by gas liquid chromatography of the products of growth, with special reference to the identification of breakdown products of glucose, has a role in the identification of *Bacteroides* organisms. (We lack experience of this approach and the reader is referred to Chapter 3).

At best, these techniques will distinguish between different species. In clinical work, it may be necessary to distinguish between different strains of the same species, to show for example whether a given infection has been exogenously or endogenously acquired, or to shed light on the number of different strains present at a given site. This is achieved with *Pseudomonas* organisms, for example, by bacteriocine typing and it might be possible to develop a similar system for the labelling of *Bacteroides* strains.

Bacteriocine-like activity in cultures of *Bacteroides* species was recently detected in our laboratories and we acknowledge the prior report of a similar finding by Beerens and Baron (1964) with organisms that we would now classify as *Bacteroides* spp. Results of our further intensive studies indicate that this may offer a new approach to the difficult problem of characterisation and subclassification of these organisms (Collee, Watt & Dewhurst, *to be published*). We have found many strains that produce these bacteriocine-like effects; a large proportion of the strains show some degree of sensitivity that is demonstrable on blood agar plates by a procedure analogous to that developed by Darrell and Wahba (1964) and by Gillies and Govan (1966) for typing *Pseudomonas* spp. (see Fig.2.5). The effect that we have observed seems to be specific for the *Bacteroides* group and we suggest the term

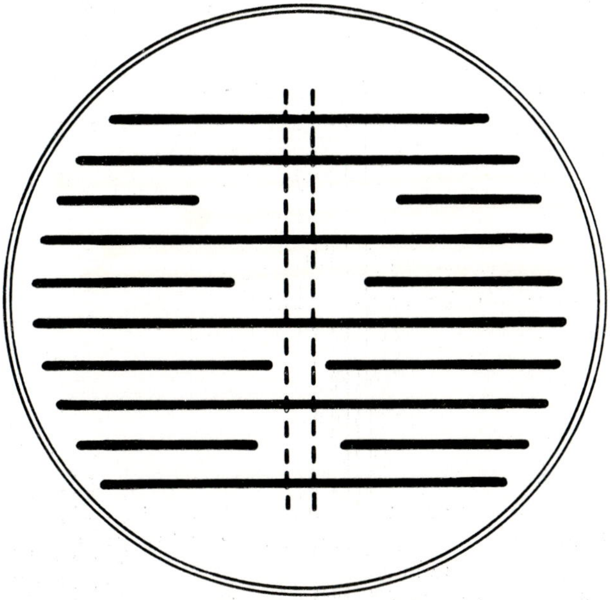

Fig.2.5. A diagram of a pattern of inhibition zones obtained by cross-streaking ten indicator strains of bacteroides-like organisms over the line of growth of a (producer) strain that has been previously grown on the medium and subsequently removed (according to the method of Gillies and Govan, 1966). Producer strains might be identified according to their specific patterns of inhibition of a standard set of indicator strains.

'bacteroidicines' for the agents involved. There are, at present, practical problems of reproducibility, and we have not yet formally established that the agents are bacteriocines, but we hope that the finding may be of practical use in the sub-classification of this difficult group and that it can be exploited for the epidemiological tracing of strains.

SUMMARY

For the isolation and identification of anaerobic bacteria of clinical interest, attention to detail at all stages is important. Consideration should be given to the prompt submission of an adequate specimen under conditions that do not inactivate bacteria; as they are at present used, swabs do not appear to meet these requirements.

In the laboratory, a standardised anaerobic jar procedure is essential. The Gaspak offers an acceptable alternative. Each of these procedures ensures that carbon dioxide is present in the anaerobic atmosphers and this is important. An anaerobic cabinet does not, at present, seem to be essential equipment.

The clinical bacteriologist is obliged to characterise isolates promptly and relatively simply. Antibiotic sensitivity testing of anaerobes and sugar fermentation tests pose special problems. In the interpretation of results, an attempt must be made to distinguish between possible contamination of the specimen with commensal flora and the presence of a disease process that may be life-threatening. In this regard, the bacteriocine-like effect that we have observed may merit further study with a view to labelling *Bacteroides* organisms.

REFERENCES

Beerens, H., & Baron, G. (1964). Mise en évidence de bactériocines élaborées par les bactéries anaérobies a Gram négatif appartenant au genre Eggerthella. *Annales de L'Institut Pasteur (Paris)*, 105, 255.

Bowden, G.H., & Hardie, J.M. (1971). Anaerobic organisms from the human mouth *in Isolation of Anaerobes*, Ed. by Shapton, D.A., & Board, R.G., London and New York, pp.177-205.

Cary, S.G., & Blair, E.B. (1964). New transport medium for shipment of clinical specimens. I. Fecal specimens. *Journal of Bacteriology*, 88, 96.

Collee, J.G., & Watt, B. (1972). Perspectives in clinical anaerobic bacteriology - where do we do from here? *in Proceedings of Becton Dickinson Symposium on 'Modern Trends in Microbiology'*, United States Trade Center, London: September, 1972.

Collee, J.G., Watt, B., Fowler, E.B., & Brown, R. (1972). An evaluation of the GasPak system in the culture of anaerobic bacteria. *Journal of applied Bacteriology*, 35, 71.

Collee, J.G. (1974). In *Medical Microbiology*. Ed. by Cruickshank, R., Duguid, J.P., Marmion, B.P., & Swain, R.H.A. Edinburgh and London, chapter 36, *in press*.

Collee, J.G., Watt, B., Brown, R. & Johnstone, S. (1974). The recovery of anaerobic bacteria from swabs. *Journal of Hygiene, Cambridge*, 72, 1.

Darrell, J.H. & Wahba, A.H. (1964). Pyocine typing of hospital strains of *Pseudomonas pyocyanea*. *Journal of Clinical Pathology*, 17, 236.

Finegold, S.M., Sugihara, P.T., & Sutter, V.L. (1971). Use of selective media for isolation of anaerobes from humans *in Isolation of anaerobes* (Society for Applied Bacteriology Technical Series No.5) Ed. by Shapton, D.A., and Board, R.G. London and New York, pp. 99-108.

Gillies, R.R., & Govan, J.R.W. (1966). Typing of *Pseudomonas pyocyanea* by pyocine production. *Journal of Pathology and Bacteriology*, 91, 339.

Moore, W.B. (1968). Solidified media suitable for the cultivation of *Clostridium novyi* type B. *Journal of General Microbiology*, 53, 415.

Stuart, R.D. (1956). Transport problems in public health bacteriology. The use of transport media and other devices to maintain the viability of bacteria in specimens. *Canadian Journal of Public Health*, 47, 114.

Watt, B. (1972*a*). Studies in bacterial anaerobiosis: the recovery of clinically important anaerobes on solid media. M.D. Thesis, University of Edinburgh.

Watt, B. (1972*b*). The recovery of clinically important anaerobes on solid media. *Journal of Medical Microbiology*, 5, 211.

Watt, B. (1973). The influence of carbon dioxide on the growth of obligate and facultative anaerobes on solid media. *Journal of Medical Microbiology*, 6, 307.

Watt, B., Hoare, M.V., & Collee, J.G. (1973). Some variables affecting the recovery of anaerobic bacteria: a quantitative study. *Journal of General Microbiology*, 77, 447.

Watt, B., Collee, J.G., & Brown, R. (1974). The isolation of strict anaerobes: the use of an anaerobic cabinet compared with a conventional procedure. *Journal of Medical Microbiology*, 7, *in press*.

3. The Isolation and Identification of Non-sporing Anaerobes

B. DRASAR

Very many methods have been described for the cultivation of anaerobic bacteria but most are now only of historic interest. Those interested in the development of anaerobic methodology should refer to the reviews of Hall (1929) and Sonnenwirth (1972); practical details of many useful techniques are presented by Shapton and Board (1971). The purpose of this paper is to discuss some techniques that have been found useful for the cultivation and identification of anaerobic bacteria.

CULTURE OF ANAEROBIC BACTERIA

Oxygen, dissolved in culture media prepared and stored aerobically, may prevent the growth of anaerobic bacteria by causing the oxidation of organic substances in the medium, by raising the oxidation-reduction potential of the medium, by inhibition of oxygen sensitive enzymes, by the destruction of oxygen sensitive compounds containing disulphide bridges, and by the diversion of metabolism to produce toxic metabolites such as hydrogen peroxide. Individual bacteria separated from one another by the plating out of specimens on media containing dissolved oxygen are maximally exposed to these toxic influences.

Dissolved oxygen can be removed from media by boiling or by allowing time for equilibrium with an oxygen-free atmosphere, and some of the effects of dissolved oxygen can be overcome by the addition of reducing agents to the culture media. Anaerobic systems utilise one or more of these techniques to obtain an anaerobic environment.

ANAEROBIC JAR

The anaerobic jar is the device most widely used to maintain an oxygen-free atmosphere around bacterial cultures. Several jars are available commercially but any gas tight container that will withstand the stress of evacuation, such as a milking machine pail or a pressure barrel, can be used. Anaerobiosis is achieved by partially evacuating (-650 mm Hg) the jar and refilling with an oxygen-free gas mixture incorporating at least 10 per cent hydrogen. Any residual oxygen is combined catalytically with the hydrogen by a suitable palladium catalyst active at room temperature such as 'D Catalyst', which is available in bulk.

As an alternative to this type of evacuation replacement system, a disposable gas generator (GasPak) may be used. This system relies on the generation of excess hydrogen for the catalytic removal of all the oxygen in the jar.

CULTURE MEDIA

Any culture medium placed in an anaerobic jar containing sufficient hydrogen and an active catalyst will become anaerobic although this may take as long as eight hours. The exposure of organisms to dissolved oxygen may be reduced by the use of freshly boiled media, freshly poured plates, or by storing media in an anaerobic atmosphere until required. The incorporation in media for plate preparation, of 0.05 per cent L-cysteine and 0.03 per cent sodium formaldelyde-sulphoxalate, as reducing agents, is recommended.

THE ANAEROBIC CHAMBER

When an anaerobic jar is used for the culture of anaerobic bacteria, all manipulations are performed on the laboratory bench and the organisms are exposed to atmospheric oxygen. In order to protect specimens and cultures from the atmosphere various glove boxes enclosing an anaerobic atmosphere have been devised (Socransky, MacDonald & Sawyer, 1959; Drasar, 1967; Aranki *et al.*, 1969). These chambers make it possible to isolate larger numbers of anaerobic bacteria from faeces and intestinal contents than is possible with conventional anaerobic jar techniques.

The apparatus

The flexible plastic chamber of the type introduced by Aranki *et al* (1969) has been most widely adopted. The apparatus consists of a flexible plastic isolator (Plysu Industrial Ltd) to which an airlock is attached (Fig.3.1). The airlock can be evacuated to enable materials to be passed into and out of the cabinet without the risk of air contaminating the cabinet atmosphere. Once set up the cabinet remains anaerobic for months. The isolator is filled with a gas mixture of 10 per cent hydrogen and 90 per cent nitrogen and contains a circulating fan, 500 g of 'D catalyst', and a tray of silica gel. For the streak inoculation of plates an electrically heated loop (Trotman & Drasar, 1968) is used.

Culture media

Specially prepared pre-reduced anaerobically sterilised media, poured into plates within the cabinet can be used. However, media prepared in the normal way, transferred into the chamber and stored for 3 days before use are almost as effective (Aranki & Freter, 1972) as indeed are commercially poured blood plates treated in this way.

Fig.3.1. An anaerobic chamber.

Culture of a specimen

The standard procedure followed in this laboratory is set out here. This procedure ensures that an anaerobic atmosphere is maintained within the cabinet. Control over the atmosphere of incubation is achieved by the evacuation and refilling of the anaerobic jar.

The cabinet ready prepared contains all the necessary media reduced by exposure to the anaerobic environment of at least 3 days.

1. Check that both airlock doors are firmly closed.
2. Open the outer airlock door and place the specimen in the airlock.
3. Close the outer airlock door and evacuate the airlock.
4. Fill airlock with 10 per cent hydrogen 90 per cent nitrogen gas mixture.
5. Open inner airlock door and transfer specimen to the interior of the isolator.
6. Inoculate relevent media.
7. Pack plates into an anaerobic jar within the cabinet.
8. Transfer jar to the outside via the airlock.
9. Evacuate and refill the jar with a gas mixture 10 per cent carbon dioxide and 90 per cent hydrogen.
10. Place anaerobic jar in the incubator.

PRE-REDUCED ANAEROBICALLY STERILISED MEDIA (PRAS)

The use of tightly stoppered, individually gassed tubes of anaerobically sterilised media was introduced by Hungate (1950) for the culture of cellulolytic bacteria from the bovine rumen. In this system each culture tube is an individual anaerobic container.

Any bacteriological culture medium may be prepared in a pre-reduced form and sterilised anaerobically. Brain heart infusion and pepone yeast extract medium are common laboratory media that can usefully be prepared in this way. Standard procedures for the preparation of PRAS media and the culture of specimens are set out below.

Preparation of the media

1. Weigh dry ingredients and place in a conical flask.
2. Add water, any necessary mineral salts solutions, and resazurin solution (final concentration 1 mg/litre of medium).
3. Place polypropylene-covered magnet in flask, fit chimney to flask and place flask on heated magnetic stirrer (Fig.3.2).

Fig.3.2. Flask fitted with chimney used in preparation of PRAS media.

4. Boil medium until the resazurin turns from pink to colourless.
5. Remove flask from the heat and remove chimney. Immediately pass a stream of oxygen-free carbon dioxide into the medium. The gas flow should be sufficient to cause gentle bubbling. Oxygen-free carbon dioxide is available from the Special Gases Division of British Oxygen.
6. When the medium is cool add L-cysteine hydrochloride to a final concentration of 0.5 grams/litre and sodium formaldehyde-sulphoxalate to a final concentration of 0.3 grams/litre.
7. Adjust the pH of the medium.
8. Dispense the medium into test tubes. During dispensing, both medium and tubes are flushed with oxygen-free nitrogen. As the tube is stoppered the gas tube is withdrawn taking care to exclude air. Butyl rubber stoppers that are impermeable to oxygen are available from the Esco Rubber Company.
9. Clamp the stoppers into the tubes. A 'Foster type' tissue culture rack with a toggle clipped lid is suitable (available from Denley Instruments Ltd).
10. Autoclave the clamped tubes at 15 lbs per square inch for 15 minutes. Allow the medium to cool before releasing the racks. Discard any tubes that are aerobic as indicated by a pink colour.

Culture of specimens

1. Melt a tube of PRAS agar taking care that the stopper does not come off.
2. Prepare a roll tube, or slope.
3. Remove the stopper and insert a sterile gas cannula emitting a gentle flow of carbon dioxide into the tube (Fig.3.3).
4. Using a long loop inoculate the roll tube in a spiral from the bottom upwards. If a slope is used, streak inoculate the specimen.
5. Replace the stopper while removing the cannula, taking care to prevent the entry of air.
6. Incubate the tube at 37°C.

The gassing apparatus illustrated in Fig.3.3 is simply constructed. The V.P.I. anaerobic apparatus, available from Bellco Glass Inc., facilitates the gassing of 3 tubes at the same time and also the spiral streaking of roll tubes.

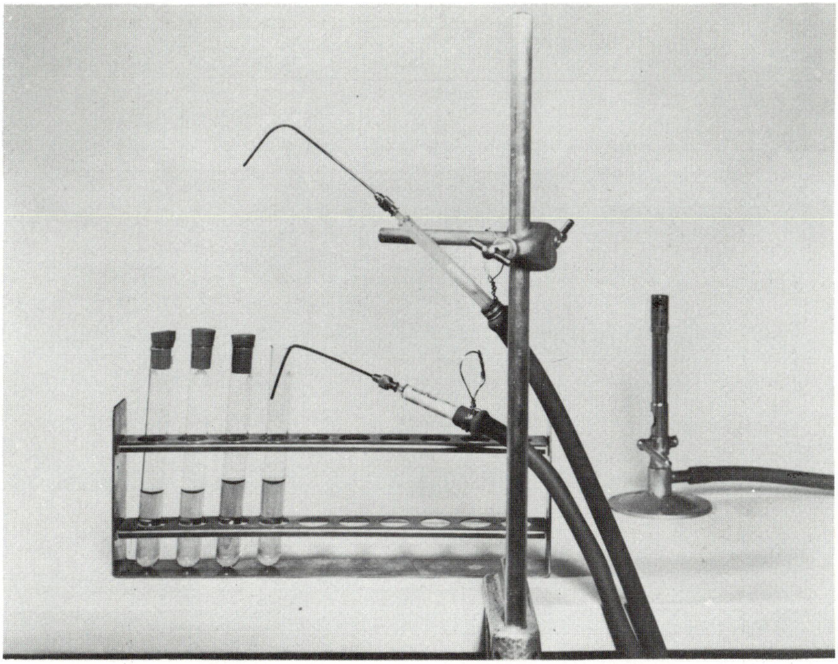

Fig.3.3. Simple gassing cannulae for inoculation and dispensing of PRAS media.

IDENTIFICATION OF NON-SPORING ANAEROBES

Anaerobic bacteria are identified on the basis of their morphology as determined by Gram's stain, fermentation reactions and metabolic tests, in the same way as the more familiar aerobic and facultative bacteria. The Anaerobe Laboratory Manual (Holdeman & Moore, 1972), provides the most complete data for the identification of the non-sporing anaerobes that may be found in infections and as part of the normal flora. A simplified scheme for use in the clinical laboratory has been proposed by Ellner, Granavo & May (1973).

END PRODUCT ANALYSIS

The classification of bacteria on a metabolic basis has long been an integral part of bacteriology. This facet of taxonomy was formalised by Kluyver & Van Neil (1936), and although effective for the differentiation of the *Enterobacteriaceae* (De Ley, 1962) such techniques have been most widely applied in studies of non-sporing anaerobes. Prevot (1940) found the identification of the acid end-products of fermentation to be useful in characterising non-sporing anaerobes, and such analysis forms a basic part of the separation of the genera in this group (Table 3.1). Various systems for the analysis of fermentation products have been described, but most of these were devised for

Table 3.1. Genera of non-sporing anaerobic bacteria represented in clinical material

Morphology	Gram Reaction	Genus	Major Fermentation Product
Rods	Gram-positive	*Propionibacterium*	Propionic acid
		Bifidobacterium	Acetic and lactic acids
		Eubacterium	Various acids
	Gram-negative	*Fusobacterium*	Butyric acid
		Bacteroides	Various acids
Cocci	Gram-positive	*Peptococcus*	Various simple patterns Main acid often acetic
		Peptostreptococcus	Various: may be very complex
	Gram-negative	*Veillonella*	Propionic and acetic acid

metabolic studies and are too laborious for routine use (Dawes *et al.* 1971). Gas chromatography has been widely adopted as the method of choice and several systems of sample preparation have been described. Steam distilation of the sample followed by injection of a concentrated distillate provides an accurate method (Chernov, Doe & Gompertz, 1972). Direct injection of acidified samples has also been used (Bricknell & Finegold, 1973). Holdeman and Moore (1972) advocate the use of ether extracts of the cultures, and two alternative methods used in this laboratory are described here.

Analysis of ether extracts of cultures

The analyses are performed with a Pye 104 gas chromatograph operated isothermally at 125°C. A flame ionisation detector is used, and the instrument is fitted with a five foot glass column 0.25 inches in diameter, packed with phosphoric acid treated 'diatomite C' and coated with 3 per cent polyethyleneglycol.

A 7 ml aliquot of a culture in glucose broth is acidified with 0.1 ml of g molar sulphuric acid. The acidified culture is extracted with 4 ml of diethyl ether. The ether extract is dried by the addition of the dry extract magnesium sulphate, and one microlitre of the dry extract is injected into the gas chromatograph. This procedure demonstrates and quantitates the volatile fatty acids.

Lactic and succinic acids are detected as their methyl-esters. To prepare these diazomethane is added to a portion of the dry ether extract, and one microlitre of the methylated extract is injected into the gas chromatograph.

Head space analysis for volatile fatty acids

The analyses are performed on a Perkin Elmer F.40 automatic head space analyser. The chromatograph operates at 130°C and the water bath at 90°C. A 2 metre long, 0.125 inch diameter column packed with Chromosorb W treated with 2 per cent phosphoric acid and coated with 15 per cent Carbowax 20 m is used for this analysis.

A 5 ml aliquot of a glucose broth culture is pipetted into the vial provided and 0.1 ml of 9 molar suphuric acid added. The vial is closed with a 20 mm butyl rubber serum stopper and the stopper sealed in position with an aluminium ring. The sealed vial is placed in the water bath; 30 vials may be loaded at one time. The vials are left for half an hour to equilibriate. When the machine is started it automatically and sequentially samples and analyses the vapor phase of each vial.

FERMENTATION REACTIONS AND BIOCHEMICAL TESTING

The metabolic tests used for the identification of aerobic bacteria can also be applied to anaerobic organisms provided that they are carried out under suitable conditions. If conventional media are used a large inoculum is often necessary to ensure adequate growth and extremely oxygen-sensitive bacteria can only be tested in PRAS sugar media. The choice of the basal medium for biochemical tests is especially important because of the complex growth requirements of some anaerobic species. Furthermore, many anaerobes alter pH indicators either by reducing the Eh of the medium or by destroying the indicator dye.

Pepone water sugars

Conventional peptone water sugars or peptone yeast extract sugars can be used for the identification of anaerobes, but they should be boiled and cooled immediately before inoculation. Incubation should be in an anaerobic jar. If a jar is not available a sterile iron nail or iron filings may be added to the medium and the bottles tightly closed.

Cooked meat medium

Media such as 'Robertson's bullock-heart medium' are employed in many laboratories for the culture of sporing anaerobes or for the enrichment of anaerobic organisms in clinical specimens. Stone (1940) used cooked meat broth as the basal medium in fermentation tests.

To ensure good growth of most anaerobes cooked meat medium prepared from fresh meat should be used. Dried preparations are inadequate to support the growth of most non-sporing anaerobes. The cooked meat medium is steamed and allowed to cool, and a sterile solution of the sugar in distilled water is added aseptically to final concentration of 1 per cent sugar in the broth. With a standard Robertson's cooked meat broth in a 30 ml universal container, 1 or 2 ml of a 10 per cent sugar solution is used. The mixed broth is then inoculated and incubated. Fermentation is detected either by the addition of indicator to the medium or by the withdrawal and testing the pH of an aliquot of the broth. Indole and hydrogen sulphide production can also be tested in this medium: indole production is best detected by the addition of Kovac's reagent to an aliquot of the broth, and a lead acetate strip is used to detect hydrogen sulphide production. The proteolysis seen in cooked meat medium should also be noted.

Thioglycollate broth

Semisolid media containing sodium thioglycollate as a reducing agent were introduced by Brewer (1940), and a similar medium containing cysteine was described by Lewis & Rettger (1940) for the examination of fermentation reactions. Fermentation media based on thioglycollate were used by Loesche, Socransky & Gibbons (1964) and Dowell & Hawkins (1968). Suitable thioglycollate media without glucose or an Eh indicator are available commercially, but the addition of 0.05 per cent cysteine hydrocloride and 0.03 per cent sodium formaldehydesulphoxalate in addition to the 0.05 per cent sodium thioglycollate is recommended.

PRAS broth

Any identification medium can be prepared in a pre-reduced anaerobically sterilised form. Broths are inoculated and incubated and after a suitable time may be tested for pH change, production of indole or any other product.

Micro-methods

1. API. This multi-test microtube system is available commercially for the identification of *Enterobacteriaceae* and lactobacilli. Results with an experimental system for the identification of anaerobes are discussed by Starr *et al.* (1973). Catteau *et al.* (1973) used the *Lactobacillus* system for the identification of bifido-bacteria. The test cards were incubated anaerobically either in a cabinet incubator or an anaerobic jar. For sugar fermentations reactions our results obtained

using this system were in agreement with more conventional systems.

2. Microtiter, Autotiter. Culture media dispensed into the disposable multi-welled plastic trays available for serological tests and tissue culture, can be used for antibiotic assay (Tilton, Lieberman & Gerlach, 1973); the trays also make ideal vessels for micro-fermentation and biochemical tests (Jayne-Williams, 1973; Huhtanen, Naghski & Dellamonica, 1972). They are sterilised under ultra-violet light, and culture media are dispensed aseptically into the wells with a dropping pipette. A drop of inoculum is then added to each well and indicator solutions and test reagents are added to the wells after anaerobic incubation.

Decarboxylase reactions

The decarboxylation of amino-acids by non-sporing anaerobes has seldom been examined. Loesche *et al.* (1964) used tests devised for the study of *Enterobacteriaceae* but in view of the changes that take place in indicator systems during anaerobic incubation, decarboxylation should be confirmed by analysis of the broth culture by thin layer chromatography.

Antibiotic sensitivity patterns

Antibiotic sensitivity patterns obtained with specially prepared, high concentration discs (Mast Laboratories) make it possible to subdivide the Gram-negative anaerobes (Sutter & Finegold, 1971). These presumptive identifications may be of value in the clinical laboratory (see p.51).

DISCUSSION

The current interest in the isolation of non-sporing anaerobes from infection was high-lighted when Moore, Cato & Holdeman (1969) reported the isolation of non-sporing anaerobes from more than 85 per cent of clinical specimens by the use of PRAS media. Their data (Table 3.2) suggest that a wide variety of anaerobic bacteria representing very many different biological types can be isolated from clinical material. If the role of these organisms in disease processes is to be assessed they must be isolated and identified.

Table 3.2. The variety of anaerobes isolated from clinical material*

Family	Genus	Number of species: Described	Number of species: Isolated from clinical material
Bacteroidaceae	*Bacteroides*	27	19
	Fusobacterium	14	9
Micrococcaceae	*Peptococcus*	7	6
Neisseriaceae	*Veillonella*	2	2
Lactobacillaceae	*Bifidobacterium*	16	5
	Peptostreptococcus	5	5
Propionibacteriaceae	*Propionibacterium*	9	7
	Eubacterium	18	10

* From the data of Moore & Holdeman (1972).

Comparison of culture systems

Man and other animals carry large numbers of anaerobic bacteria, many of them extremely oxygen-sensitive. Some of these organisms cannot be isolated by anaerobic jar techniques; for the study of these organisms the use of a PRAS system or an aerobic cabinet is essential (e.g. Hungate, 1950; Sutter *et al.* 1972).

Some anaerobic bacteria can be isolated from clinical specimens no matter which anaerobic system is used but no study has evaluated the influence of the many factors involved. However, by pooling the data from several studies, some estimates can be made (Table 3.3).

Carbon dioxide is known to be important for the growth anaerobic bacteria (Watt, 1973). Comparison of the results of Vargo *et al.* (1971) with those of Peach & Hayek (1974) suggests that the rate of isolation of anaerobes from routine specimens can be improved by about 15 per cent by the use of carbon dioxide, but that a further 20 per cent improvement results from the use of an anaerobic cabinet. If the results of the study of specimens from anaerobic infections are considered the method used for the collection and transport of specimens is seen to be crucial (McMinn & Crawford, 1970; Rosenblatt, Fallon & Finegold, 1973). If specimens are collected and transported anaerobically, all culture systems seem equally satisfactory. However, it must be noted that the results obtained with PRAS and anaerobic cabinet systems are not affected by the previous handling of the specimen. In none of the studies cited above were extremely oxygen-sensitive bacteria of the type found in the normal flora isolated. As these organisms cannot be isolated by anaerobic

Table 3.3. Comparison of the results with various anaerobic culture systems.

Specimens and lesions	Anaerobic collection and transport	Anaerobic System	At least 5% carbon dioxide atmosphere	Number of specimens examined	Percentage of specimens yielding non-sporing anaerobes	Reference
Wounds and	-	Anaerobe jar (G)*	+	145	18	Vargo *et al.* (1971)
abscesses	+	Anaerobe cabinet	+		41	
Wound	-	Anaerobe jar	-	64	4.5	Peach & Hayek (1974)
swabs	-	Anaerobe cabinet	+		42	
Anaerobic	-	Anaerobe jar (G)*	+	38	8.2	McMinn & Crawford (1970)
infections	-	PRAS	+		100	
Anaerobic	+	Anaerobe jar	+	23	100	Rosenblatt *et al.* (1973)
infections	+	Anaerobe cabinet	+		100	
	+	PRAS	+		100	

(G)* - Gaspak system used.

jar techniques, and in view of the improved isolation obtained from routinely handled specimens using PRAS and anaerobic cabinet systems, it is suggested that such systems should be adopted routinely at least until the role of very oxygen-sensitive non-sporing anaerobes in infection is more closely defined.

THE IDENTIFICATION OF NON-SPORING ANAEROBES

Reliable differentiation of the genera of non-sporing anaerobes can only be achieved if the end-products of fermentation are analysed (Table 3.1). End-product analysis is especially important for differentiation among the Gram-positive non-sporing anaerobes. If an attempt is to be made to identify these bacteria access to a gas chromatograph is essential.

The technique used for isolation determines the system used for the biochemical testing of the isolates. Thus, bacteria isolated in an anaerobic jar will grow well in both PRAS and conventional broths, but bacteria isolated by means of PRAS or cabinet techniques may not grow in conventional broths.

The use of cooked meat broth as a fermentation base is a useful emergency procedure. Thioglycollate sugars will grow virtually all anaerobic bacteria apart from very oxygen sensitive organisms. Provided their limitations are realised, these sugars are a valuable adjunct to the biochemical testing of anaerobic bacteria isolated in the clinical laboratory.

PRAS sugars have been used in most recent taxonomic studies of non-sporing anaerobes. If a laboratory is particularly interested in the classification of non-sporing anaerobes this system should be considered. If an anaerobic cabinet or the PRAS identification system are adopted for the isolation of extremely oxygen-sensitive anaerobes from clinical sources, the use of PRAS sugar media is essential. The API system requires further development for routine use with anaerobic bacteria. Microfermentation tests using microtiter trays were developed for use in ecological studies where it is essential to examine large numbers of strains. With these techniques one hundred strains can be examined daily.

Although it may be desirable to identify all anaerobic bacteria that have been isolated, especially if an improved isolation procedure is adopted, the system used for biochemical testing will be determined by the circumstances of the individual laboratory.

REFERENCES

Aranki, A. & Freter, R. (1972). Use of anaerobic glove boxes for the cultivation of strictly anaerobic bacteria. *American Journal of Clinical Nutrition*, 25, 1329.

Aranki, A., Syed, S.A., Kenney, F.B. & Freter, R. (1969). Isolation of anaerobic bacteria from human gingiva and mouse caecum by means of a simplified glove-box procedure. *Applied Microbiology*, 17, 568.

Brewer, J.H. (1940). Clear mediums for the 'aerobic' culture of anaerobes. *Journal of the American Medical Association*. 115, 598.

Bricknell, K.S. & Finegold, S.M. (1973). A simple, rapid method to process and assay fatty acids and alcohols by gas chromatography. *Analytical Biochemistry*, 51, 23.

Catteau, U., Poncelet, F., Beerens, H. & Losfeld, J. (1973). A rapid method for the identifications of *Bifidobacterium* species using 50 characters. In *Actinomycetales*. Ed. Sykes, G. & Skinner, F.A. page 301, Academic Press, London.

Chernov, A.J., Doe, W.F. & Gompertz, D. (1972). Intrajejunal volatile fatty acids in the stagnant loop syndrome. *Gut*, 13, 103.

Dawes, E.A., McGill, D.J. & Midgley, M. (1971). Analysis of Fermentation Products. In *Methods in Microbiology 6A* Ed. Norris, J.R. & Ribbous, D.W. p.53-215. Academic Press, London.

De Ley, J. (1962). Comparative biochemistry and Enzymology in bacterial classification. In *Microbiol Classification*. p.164. Cambridge University Press.

Dowell, V.R. & Hawkins, T.M. (1968). Laboratory Methods in Anaerobic Bacteriology. Publication No.1803 U.S. Government Printing Office, Washington DC.

Drasar, B.S. (1967). Cultivation of anaerobic intestinal bacteria. *Journal of Pathology and Bacteriology*, 94, 417.

Ellner, P.D., Granato, P.A. & May, G.B. (1973). Recovery and identification of anaerobes: A system suitable for the routine clinical laboratory. *Applied Microbiology*, 26, 904.

Hall, I.C.. (1929). A review of the development and application of physical and chemical principles in the cultivation of obligately anaerobic bacteria. *Journal of Bacteriology*, 17, 255.

Holdeman, L.V. & Moore, W.E.C. (1972). Anaerobe Laboratory Manual, Virginia Polytechnic Institute and State University Anaerobe Laboratory, Blacksburg, Virginia 24060.

Huhtanen, C.N., Naghski, J. & Dellamonica, E.S. (1972). Microfermentation series for identification of single colonies of Enterobacteriaceae. *Applied Microbiology*, 24, 618.

Hungate, R.E. (1950). The anaerobic mesophilic cellulolytic bacteria. *Bacteriological Reviews*, 14, 1.

Jayne-Williams, D.J. (1973). *Journal of Applied Bacteriology* in press.
Kluyver, A.J. & Van Niel, C.B. (1936). Prospects for a natural system of classification. *Zentralblatt fur Bakteriologie II Orig.*, 94, 369.
Lewis, K.H. & Rettger, L.F. (1940). Non-sporulating Anaerobic bacteria of the intestinal tract. *Journal of Bacteriology*, 40, 287.
Loesche, W.J., Socransky, S.S. & Gibbons, R.J. (1964). *Bacteroides oralis* proposed new species isolated from the oral cavity of man. *Journal of Bacteriology*, 88, 1329.
McMinn, M.T. & Crawford, J.J. (1970). Recovery of anaerobic micro-organisms from clinical specimens in pre-reduced media versus recovery by routine clinical laboratory methods. *Applied Microbiology*, 19, 207.
Moore, W.E.C., Cato, E.P. & Holdeman, L.V. (1969). Anaerobic bacteria of the gastrointestinal flora and their occurrence in clinical infections. *Journal of Infectious Diseases*, 119, 641.
Peach, S. & Hayek, L. (1974). In preparation.
Prevot, A.R. (1940). Manuel de classification des Anaerobies. Masson et Cie, Paris.
Rosenblatt, J.E., Fallon, A. & Finegold, S.M. (1973). Comparison of methods for isolation of anaerobic bacteria from clinical specimens. *Applied Microbiology*, 25, 77.
Shapton, D.A. & Board, R.G. (1971). Isolation of Anaerobes, Academic Press, London.
Socransky, S., MacDonald, J.B. & Sawyer, S. (1959). The cultivation of *Treponema microdentium* as surface colonies. *Archives of Oral Biology*, 1, 171.
Sonnenwirth, A.C. (1972). Evolution of anaerobic methodology. *American Journal of Clinical Nutrition*, 25, 1295.
Starr, S.E., Thompson, F.S., Dowell, V.R. & Balows, A. (1973). Micromethod system for identification of anaerobic bacteria. *Applied Microbiology*, 25, 713.
Stone, M.L. (1940). Studies on the anaerobic streptococcus. *Journal of Bacteriology*, 39, 559.
Sutter, V.L., Attebury, H.R., Rosenblatt, J.E., Bricknell, K.S. & Finegold, S.M. (1972). Anaerobic Bacteriology Manual. The Anaerobic Bacteriology Laboratory, Wadsworth Hospital Center, Los Angeles, California 90073.
Sutter, V.L. & Finegold, S.M. (1971). Antibiotic disc susceptibility test for rapid presumptive identification of gram-negative anaerobic bacilli. *Applied Microbiology*, 21, 13.
Tilton, R.C., Lieberman, L. & Gerlach, E.H. (1973). Microdilution antibiotic susceptibility test. Examination of certain variables. *Applied Microbiology*, 26, 658.
Trotman, R.E. & Drasar, B.S. (1968). Electrically heated inoculating loop. *Journal of Clinical Pathology*, 21, 224.

Vargo, V., Michaelson, T.C., Spaulding, E.H., Vitagliano, R., Swenson, R.M. & Forsch, E. (1971). Comparison of a pre-reduced anaerobic method and GasPak for isolating anaerobic bacteria. *Bacteriological Proceedings*, p.109.
Watt, B. (1973). The influence of carbon dioxide on the growth of obligate and facultative anaerobes on solid media. *Journal of Medical Microbiology*, 6, 307.

DISCUSSION

DR SPENCER (FRENCHAY HOSPITAL, BRISTOL): Is there any advantage of 10 per cent CO_2 over the 2 per cent that we are normally using?
DR WATT: We have shown that although 10 per cent isn't a magic number, 5 per cent is probably the lower limit of CO_2 that is necessary particularly for the non-sporing anaerobes. Ten per cent is a convenient figure because if it goes down to 8 per cent it doesn't matter, but if you start at 5 per cent and it decreases, it's down to 2 or 3 per cent and that is perhaps too little.
DR GOODWIN (NORTHWICK PARK HOSPITAL): In the isolation of organisms from the upper bowel, have the speakers had experience of the use, as I believe they do in Australia, of dropping the liquid straight into a situation where they are frozen, for instance into solid CO_2? The anaerobes apparently survive better than if the container is transported in ice.
DR DRASAR: In our studies on small and large bowel we used a glycerol transport medium to protect organisms during rapid freezing in CO_2 -acetone mixtures or liquid Nitrogen and we found it to be effective.
DR SELWYN (WESTMINSTER HOSPITAL): Could Dr Drasar elaborate a little on those extremely fastidious, awkward organisms that are found on the surface of the bowel wall? Recently I was told to my amazement that most of these are not *Bacteroides* or anything to do with the Gram-negatives, they are really very aberrant clostridia and can be found to form spores under extraordinary conditions. These by the way are seen wonderfully in electron microscopic sections lying as thick mats.
DR DRASAR: Dr Finegold has, I think, devoted more study to these organisms than anyone else and as I remember his work he found that many of them were Gram-positive cocci; it is our experience that these are very oxygen-sensitive.
DR GHONEIM (EXETER): I want to ask Dr Watt about the possibility of doing biochemical reactions in an anaerobic

jar; what is the concentration of blood used for the isolation of the anaerobes; and is there any possibility of using antibiotics such as gentamicin in selective media?

DR WATT: Certainly it is possible to do biochemical tests such as carbohydrate fementation tests in an aerobic jar as long as one has suitable controls. As far as the concentration of blood is concerned, for normal purposes we use horse blood at 10 per cent and human blood at 5 or 10 per cent. With some of the very difficult organisms if you use a lot of blood, up to 40 per cent you would appear to get better results but it's not very practical for a routine laboratory and it is fair to say that 10 per cent is a convenient figure for human or horse blood. As far as antibiotics are concerned, Dr Finegold and his colleagues have worked a lot on this and I hesitate to answer. Antibiotics can be used for selection by incorporating them in the medium, and we particularly use neomycin blood agar. Alternatively you can put disks on the surface of the plate and from the inhibition patterns a clear picture of the organism is obtained. There is the added bonus that you inhibit any other organisms that are present and this allows you to see the anaerobes clearly.

DR COLLEE: The concentration of neomycin at Edinburgh is 70 µg/ml in the final dilution.

DR PERCIVAL: Just two points: one about selective media and that is that many anaerobes, certainly *Bacteroides*, occur in mixed infections with enterobacteria and may be masked unless you use selective media. If you look at the literature you will find that the highest percentage isolation rates from clinical specimens have been where people have used both selective media and enrichment; I think this is very necessary in the ordinary clinical lab. The next point I would like to stress is the question of the importance of fresh specimens. We did a study where we took swabs at the bedside in duplicate and plated them out either directly or allowed them to go to the routine laboratory through the normal channels. We had a 30 per cent isolation rate from those that went through the usual channels, taking up to 18 hours, but a 70 per cent isolation rate from the same patients when the specimens were plated out by the bedside. I think you can say in people who have had their bowel opened you can expect to find *Bacteroides* in great profusion in very nearly all of the wounds that contain enterobacteria.

DR COLLEE: Thank you Dr Percival. We tried to summarize this in one of our papers some time ago by saying that a doctor's responsibility to kill organisms in the interest of his patient shouldn't extend along the route to the laboratory!

4. Antibiotic Sensitivity of Non-sporing Anaerobes

I. PHILLIPS

One of the consequences of the increasing awareness of the importance of anaerobic non-sporing organisms in infectious processes, has been the increasing interest in the assessment of the antimicrobial susceptibility of these organisms. Most of the work has been with *Bacteroides fragilis*, a reflection of the undoubted clinical importance of this organism, but also a reflection of the greater difficulties in isolating and maintaining many of the others. This paper reviews the work that has been done so far, and presents new data on *B. fragilis*.

METHODS OF ASSESSING ANTIMICROBIAL SUSCEPTIBILITY OF ANAEROBES

In 1955, Garrod reviewed the earlier limited literature on the determination of minimum inhibitory concentrations (MIC's) of antibiotics for *Bacteroides*, and in addition to establishing characteristic patterns of susceptibility for the *Bacteroides* species then recognised, he established that the agar dilution method of determining MIC's was applicable to anaerobes (Garrod, 1955). Gillespie & Guy (1956) and Finegold & Hewitt (1956) came to similar conclusions. Later work has confirmed that the agar dilution method and the tube dilution method are valid and has extended their scope to include Gram positive as well as Gram negative anaerobes (Ingham, Selkon, Codd & Hale, 1968; Thornton and Cramer, 1971; Finegold & Sutter, 1972; Kislak, 1972; Martin, Gardner, & Washington, 1972). Finegold & Sutter (1972) have underlined the necessity for standardisation of techniques if results are to be comparable. A major problem, among many others, is the effect of CO_2, presence of which renders erythromycin and lincomycin MIC's up to 32 times higher than those determined in its absence (Ingham, Selkon, Codd & Hale, 1970).

Minimum bactericidal concentrations (MBC's) of a variety of agents have also been determined. Some workers have used the standard method of subculture from tube dilution tests (Martin *et al.*, 1972; Nastro & Finegold, 1972) while others have used replica plating from solid media (Ingham *et al.*, 1968; Whelan & Hale, 1973).

The many problems, in particular that of inoculum size, that beset the assessment of antibiotic susceptibility by disc methods, have recently received considerable publicity. To these are added, in the present context, problems related to anaerobiosis itself, as well as to the effects of CO_2, pH of media, presence of blood, reducing substances and other materials, and duration of incubation (Rosenblatt & Schoenknecht, 1972). It is, in addition, essential to use an anaerobic organism as a control in these tests (Finegold & Sutter, 1972). However Dr Finegold and his colleagues and others have shown that provided a standardised method is used, reliable results can be obtained for many antibiotics, and many species of micro-organisms (Sutter, Kwok & Finegold, 1972, 1973; Bodner, Koenig, Treanor & Goodman, 1972). In a study of *B. fragilis* it was observed that susceptibility to chloramphenicol, clindamycin, penicillin, vancomycin, and tetracycline could be determined by a standard disc method but that results for erythromycin and lincomycin were less reliable (Sutter *et al.*, 1973). Because of a lack of attention to the problems of interpretation of disc susceptibility tests, and a failure to correlate results with MIC's, many reports on antibiotic susceptibility assessed by this method are difficult to evaluate. Figures 4.1 - 4.4 show our preliminary attempts at the correlation of results of disc tests and MIC determinations for four antibiotics and *B. fragilis*. In our hands,

---

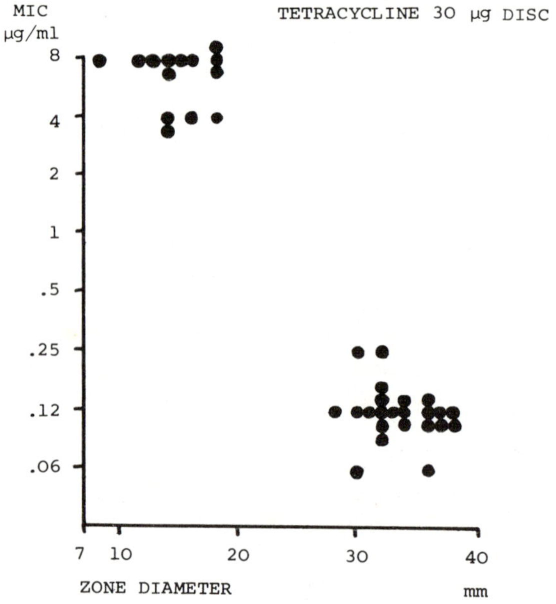

Fig.4.1. Correlation of results of disc susceptibility tests and MIC's for tetracycline and *B. fragilis*.

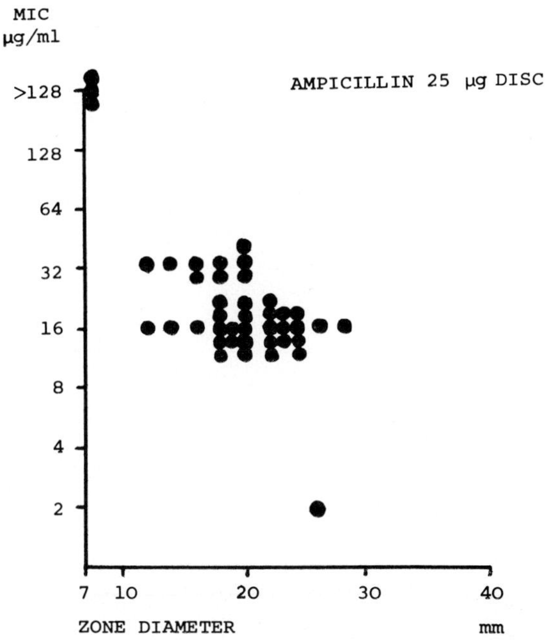

Fig. 4.2. Correlation of results of disc susceptibility tests and MIC's for ampicillin and *B. fragilis*.

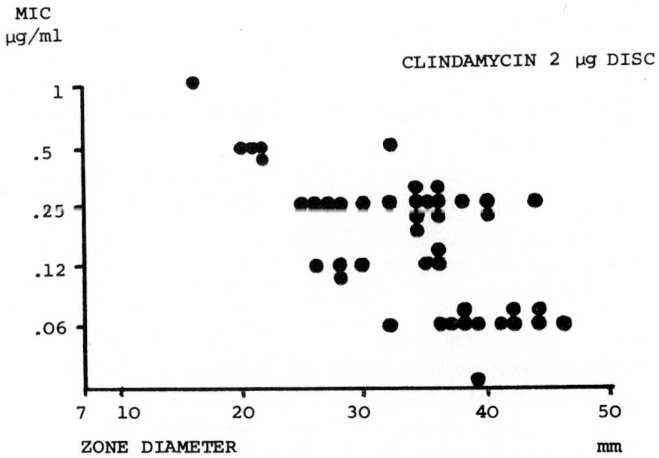

Fig. 4.3. Correlation of results of disc susceptibility tests and MIC's for clindamycin and *B. fragilis*.

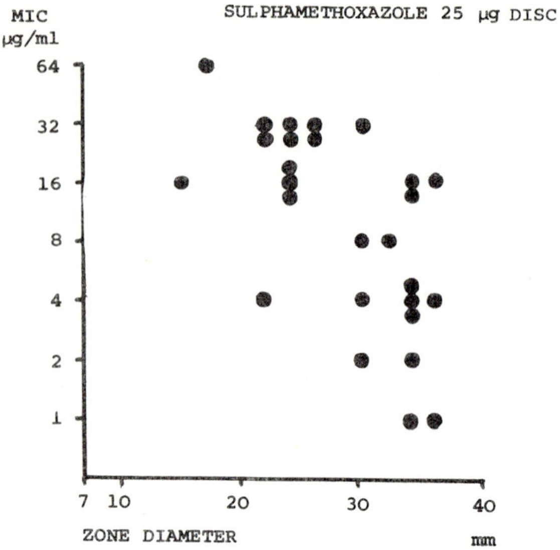

Fig. 4.4. Correlation of results of disc susceptibility tests and MIC's for sulphamethoxazole and *B. fragilis*.

the disc tests reliably distinguish strains highly resistant to ampicillin (Fig.4.1), or moderately resistant to tetracycline (Fig.4.2), but are less clear-cut for clindamycin (Fig.4.3) and for sulphamethoxazole (Fig.4.4).

ANTIMICROBIAL SUSCEPTIBILITY OF ANAEROBES

A recent review of the antimicrobial susceptibility of anaerobic Gram-negative bacilli by Finegold & Sutter (1972) includes an excellent bibliography on the whole field of Gram-positive as well as Gram-negative anaerobes, and includes a useful summary of the work of Dr Finegold and his colleagues at the Wadsworth General Hospital, Los Angeles. An additional important source of information is a report on a study of the susceptibility of some 600 clinical isolates of a variety of anaerobes to 10 antimicrobial agents, by Martin *et al.*, (1972) at the Mayo Clinic.

ANAEROBIC COCCI

Table 4.1 shows the characteristic antimicrobial susceptibility of *Peptococcus*, *Peptostreptococcus*, and *Veillonella* species. It should be noted that very few strains of *Veillonella* have been investigated. All three genera resemble each other, and many aerobic cocci, in being almost invariably susceptible to benzylpenicillin, chloramphenicol, clindamycin, and rifampicin, but usually resistant to aminoglycosides, and sometimes to tetracycline.

Table 4.1. Usual MIC's for *anaerobic cocci*.

	Peptococcus	*Peptostreptococcus*	*Veillonella*
benzylpenicillin	.1 to 3.1*	.1 to 1.6	.2
chloramphenicol	3.1*	3.1	1.6
clindamycin	.1 to 6.2*	.1 to 1.6	.1
rifampicin	.1 to 6.2*	.1 to 6.2*	.1 to 1.6
tetracycline	.1 to R	.1 to R	.1 to 25
erythromycin	.1 to R	.1 to 12.5	.2 to R
metronidazole	<6.2*	6.2 to R	<6.2*
gentamicin	3.1 to 25*	.8 to R	6.2 to R
kanamycin	1.6 to R	1.6 to R	.1 to R

* - occasional more resistant strains R resistant

after Martin *et al.*, (1972), Tally, Sutter & Finegold (1972), Fass *et al.*, (1973).

They differ from aerobes in their lower susceptibility to erythromycin, but in this context some of the technical problems that have already been mentioned are particularly difficult. They quite clearly differ from aerobes and resemble other anaerobes in their susceptibility to metronidazole. There is no published information about the bactericidal activity of any of these compounds on bacteria in this group.

ANAEROBIC NON-SPORING GRAM-POSITIVE BACILLI

The results of determinations of MIC's of a variety of agents for *Propionibacterium acnes*, *Eubacterium*, *Bifidobacterium*, *Catenabacterium*, and *Actinomyces*, by a variety of authors are summarised in Table 4.2. It should be emphasised again that results are often based on the examination of small numbers of strains. Once again, the organisms resemble aerobic Gram-positive species in their high susceptibility to benzylpenicillin, chloramphenicol, clindamycin, and rifampicin, their resistance to aminoglycosides, and their varying susceptibility to tetracycline. Like anaerobic cocci they are not infrequently resistant to erythromycin. *Bifidobacterium* species, like other anaerobes are susceptible to metronidazole: reports on the metronidazole susceptibility of other species have yet to appear. It is also known that bifidobacteria are susceptible to vancomycin, and that *Actinomyces israelii* is susceptible to fusidic acid. It would be of considerable interest to fill in the gaps in our knowledge of the susceptibility of these organisms to such agents as fusidic

Table 4.2. Usual MIC's for anaerobic *Gram-positive non-sporing bacilli*

	Propionibacterium acnes	*Eubacterium*	*Bifidobacterium*	*Catenabacterium*	*Actinomyces*
benzylpenicillin	.1	.1 to 1.6	<.2 to .4	.2	.1
tetracycline	.8	.2 to R	1.6 to R	1.6	2
chloramphenicol	1.6	.4 to 6.2	3.1	.8 to 3.1	2
erythromycin	1. to 6.2*	.2 to 12.5*	.1 to .8	.1	.1
clindamycin	.2	.4	.1	.4	<.5
gentamicin	3.1 to R	.1 to R	.8 to 25	.4 to R	.
rifampicin	.1*	.4*	.4 to 1.6	.1 to 1.6	.
fusidic acid	.	.	.	.	.5
vancomycin	.	.	.2 to 1.56*	.	.
metronidazole	.	.	6.2	.	.

R resistant * - occasional more resistant strains, . - information not available

after Martin *et al.*, (1972), Garrod, Lambert & O'Grady (1973), Miller & Finegold (1967), Tally *et al.*, (1972), Lerner (1969), Blake (1964).

acid, metronidazole, sulphonamides, and trimethoprim.

ANAEROBIC GRAM-NEGATIVE BACILLI

These organisms differ considerably from aerobic Gram-negative rods in their susceptibility to antibacterial agents, and are thus of considerable interest. *Fusobacterium* species and *Bacteroides* species have been the subject of several reports, and will be considered separately.

Fusobacterium species

Usual MIC's for four species, *F. fusiforme*, *F.necrophorum*, *F. mortiferum* and *F. varium*, are listed in Table 4.3. With the exception of the aminoglycosides, vancomycin, and erythromycin, most antibiotics are highly active against *Fusobacteria*. One important exception is rifampicin which is not active against *F. mortiferum* and *F. varium*, which therefore differ from *Bacteroides* species and other fusobacteria. *Fusobacterium* species also differ from *Bacteroides* in their almost universal sensitivity to benzylpenicillin and polymyxin B, but resemble *Bacteroides* in being almost always susceptible to clindamycin and metronidazole, but sometimes resistant to tetracycline.

Bacteroides species

Usual MIC's of a variety of agents for *B. fragilis*, *B. melaninogenicus*, *B. oralis*, *B. corrodens* and other species are given in Table 4.4. With few exceptions they are susceptible to clindamycin, metronidazole, chloramphenicol, and rifampicin, and resistant to aminoglycosides, polymyxins, and vancomycin. They are often resistant to benzylpenicillin, erythromycin, and tetracycline. *B. fragilis* is known to be usually susceptible to nitrofurantoin and according to Mitchell (1973) it is sensitive on disc testing to fusidic acid.

B. fragilis

This is possibly the most important of the anaerobes in clinical medicine, and therefore merits more detailed discussion.

It has already been stated that *B. fragilis* is often resistant to benzylpenicillin. Finegold & Sutter (1972) have suggested that ampicillin, carbenicillin, and cephaloridine are roughly equivalent to penicillin, while phenethicillin, methicillin, cephalothin, and cephalexin are less active. Kislak's results are in general agreement with this (Kislak, 1972). The degree of susceptibility to penicillin, ampicillin, and carbenicillin in particular might bring

Table 4.3. Usual MIC's for *Fusobacterium* species.

	F. fusiforme	*F. necrophorum*	*F. mortiferum*	*F. varium*
clindamycin	.1 to 1.6	≤.1	≤.1	.1 to 25
metronidazole	≤.1	≤.1	.	.4
chloramphenicol	.1 to 6.2	.8	.8	3.1
benzylpenicillin	.1*	<.1	.2 to 1.6	.1 to 3.1*
tetracycline	.1 to 6.2*	.1	.2	.1 to 25
rifampicin	.4 to 3.1	.	>100	>50
erythromycin	.1 to R	.8 to 6.2	>25	>25
gentamicin	12.5 to 400	200	200	200
kanamycin	.1 to 800	200 to 3200	200	400
vancomycin	6.2 to R	>25	>25	>25
polymyxin B	.4	.	.	.

* - occasional more resistant strains, R - resistant

after Finegold & Sutter (1972), Martin *et al.* (1972).

Table 4.4. Usual MIC's for *Bacteroides* species.

	B. fragilis	*B. melaninogenicus*	*B. oralis*	*B. corrodens*	Other *Bacteroides* species
clindamycin	.1 - 6.2*	.1 to 1.6	⩽.1	.8	⩽1.6
metronidazole	.4 - 3.1*	.8	.8 to 12.5	.8 - 6.2	.
chloramphenicol	6.2	.1 to 6.2	1.6	3.1	3.1
benzylpenicillin	12.5 to R	.1 to 25	.1 to 3.1	.4	.1 to 12.5
tetracycline	.4 to 25	.1 to 25	.4 to 6.2	.8	.1 to R
rifampicin	.4 to 1.6	.1 to .4*	.1 to 1.6	.	.1 to 1.6
erythromycin	3.1 to R	.1 to .8*	.2 to 1.6	1.6 - 12.5	.1 to 1.6*
gentamicin	25 to 1600	25 to 400	6.2 to 400	.	.1 to R
kanamycin	>6400	.1 to >6400	100 to >6400	⩽25	.2 to R
vancomycin	⩾25	1.6 to R	.	⩾25	.
polymyxin B	>100	.4 to 50	.2 to >100	.	.
nitrofurantoin	6.2	.	.	.	.

* - occasional more resistant strains, R - resistant.

after Finegold & Sutter (1973), Martin *et al.* (1972), Ingham *et al.* (1968).

infections into the treatable range if sufficiently large doses were used (Kuesch & O'Connell, 1966). Our results (Table 4.5) are similar for ampicillin and carbenicillin, with most of our strains having MIC's lower than easily attainable serum levels. There has been considerable discussion on whether resistance to penicillins and cephalosporins is due to beta-lactamase production. Garrod (1955) observed a penicillinase effect in disc susceptibility tests with two of the nine strains that he examined, and Pinkus, Veto & Braude (1968) have demonstrated penicillinase activity particularly among organisms with higher penicillin MIC's. Anderson & Sykes (1972) have studied one strain that was exceptionally resistant to beta-lactam antibiotics, and produced a unique beta-lactamase that was more active against cephaloridine than against the penicillins. Del Bene & Farrar (1973) have found that a beta-lactamase characterized by its greater activity against cephalosporins was present in varying amounts in all of ten strains of *B. fragilis* that they examined, and that there was a correlation between the amount of enzyme present and the in vitro resistance to cephaloridine. Braude (1973) found that over 90 per cent of 300 strains examined by him produced penicillinase, and outlines the technical problems involved in demonstrating its presence. We have, perhaps surprisingly found little evidence of an inoculum effect on ampicillin and carbenicillin MIC's, as might be expected if resistance were due to a beta-lactamase. Our finding of four strains of *B. fragilis*, among over 100 now examined, considerably more resistant to ampicillin and carbenicillin than other strains suggests more than a quantitative difference, but the B-lactamase (if any) has yet to be characterised.

There is general agreement that *B. fragilis* is resistant to aminoglycosides (Finegold & Sutter, 1971) and our results merely add to the already conclusive data. However we have been surprised to find that these organisms are as susceptible as gonococci to spectinomycin.

Our results for clindamycin, metronidazole, tetracycline, minocycline, erythromycin, chloramphenicol, rifampicin and fusidic acid are given in Table 4.6 (Phillips & Warren, 1974). They largely confirm those of others in the United Kingdom (Ingham *et al.*, 1968, 1970; Mitchell, 1973; Okubadejo, Green & Payne, 1973*a*, *b*). All the strains that we have examined were susceptibile to clindamycin, and all but two out of 70 were susceptible to metronidazole. It is of interest that the two metronidazole-resistant strains were also resistant to tetracycline.

Only 22 per cent of our strains were resistant to tetracycline, and none of these were among the 17 isolates

Table 4.5. *Bacteroides fragilis* MIC's of ampicillin, carbenicillin, aminoglycosides, and spectinomycin (Phillips & Warren, 1974).

	MIC 2	4	8	16	32	64	128	256	512	μg/ml 1024	Number of strains tested
ampicillin	1		8	34	15					4*	62
carbenicillin		4	19	23	9	3	1			3*	62
kanamycin							32*				32
BBK 8							32*				32
gentamicin							32*				32
tobramycin							32*				32
spectinomycin		1	2	58	7	1					69

* MIC >figure given (higher concentrations not tested)

Medium: Oxoid DST plus 6 per cent lysed horse blood. Inoculum: 10^5 organisms.

Table 4.6. *Bacteroides fragilis*: MIC's of clindamycin, metronidazole, tetracycline, minocycline, cloramphenicol, erythromycin, rifampicin, fusidic acid (Phillips & Warren, 1974).

	MIC										μg/ml	Number of
	.007	.015	.03	.06	.12	.25	.5	1	2	4	8	strains tested
clindamycin		3	3	13	11	28	6	1				65
metronidazole						7	30	31			2*	70
tetracycline			9	20	11				1	9	3*	53
minocycline	10	27			1	9	6					53
chloramphenicol								5	21	30		56
erythromycin				6	42	5					1	54
rifampicin		2	12	37	1	1						53
fusidic acid						1		17	32	3		53

* MIC >figure given (higher concentrations not tested)

Medium: Oxoid DST plus 6 per cent lysed horse blood. Inoculum: 10^5 organisms from blood cultures. Tetracycline resistance would appear to be less common in Britain than in the United States, but may be increasing. Ingham *et al.* (1968) found no tetracycline resistance in the 17 strains that they examined but more recently Okubadejo *et al.* (1973*a*) and Mitchell (1973) have found 33 per cent and 38 per cent resistant respectively. Finegold & Sutter (1972) have reported up to two-thirds of their isolates resistant to tetracycline. Like Finegold & Sutter we have found minocycline about four to eight times as active as tetracycline, and it is possible that infections with organisms relatively resistant to tetracycline might be treatable with minocycline.

Results with erythromycin are difficult to interpret. We have found that MIC's determined on plates incubated in the presence of 10 per cent CO_2 are lower than those of most workers, and that, as expected, those determined in the absence of CO_2 are lower still. Differences in technique are presumably responsible and it seems possible that the true determination of susceptibility would need in vivo tests. Our rifampicin MIC's are also lower than those of most workers. Results for fusidic acid appear to confirm Mitchell's (1973) conclusions from disc tests, that most strains of *B. fragilis* are susceptible to fusidic acid.

There is little information on the susceptibility of *B. fragilis*, or other anaerobes, to sulphonamides and trimethoprim. On the basis of tests with discs containing both agents, Okubadejo *et al.* (1973*a*) found all of 60 strains of *B. fragilis* susceptible. On the other hand Leigh (1973) implies that the organism is usually resistant. Results of MIC determinations have been published for only three strains, by Bushby (1973), who concluded that the organisms were sensitive to sulphamethoxazole (MIC 10 µg/ml) but resistant to trimethoprim (MIC 50 µg/ml) and that there was little evidence of potentiation when the two agents were mixed. Our results (Phillips & Warren, 1974) are somewhat at variance with those of Bushby. In a study of 48 strains we found that most were more resistant to trimethoprim than to sulphonamide, but that few were as resistant as those described by Bushby. When the two agents were combined in the ratio of their MIC's considerable potentiation was observed. Figure 4.5 is an isobologram of FIC's* of trimethoprim and sulphamethoxazole combined in the proportions 20 : 1(A), 1 : 1(B), and 1 : 20(C), based on means of results for 17 strains of *B. fragilis*, and a similar isobologram for the Oxford staphylococcus. The mean of the optimum proportions of trimethoprim and sulphamethoxazole for maximum potentiation of activity against *B. fragilis* lies between 20 : 1 and 1 : 1, while,

---

$$* \ FIC = \frac{\text{MIC of agent in combination}}{\text{MIC of agent alone}}$$

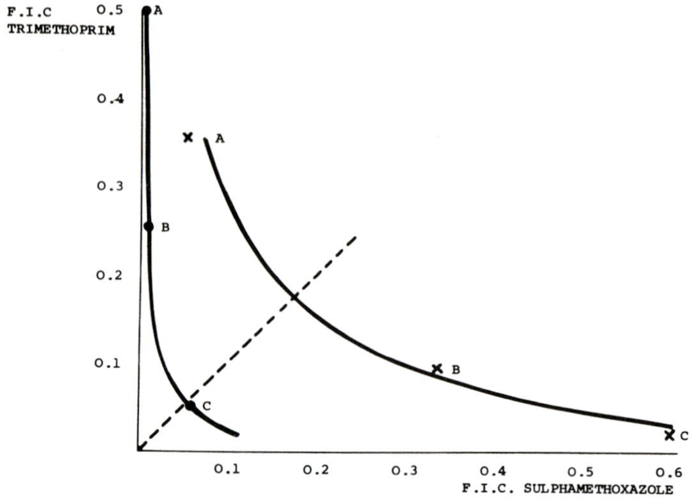

Fig. 4.5. Synergism of trimethoprim and sulphamethoxazole for *Bacteroides fragilis* and Oxford *staphylococcus*. F.I.C. = Fractional Inhibitory Concentration; X = *B. fragilis*; . = Oxford Staphylococcus; Ratio concentration trimethoprim : sulphamethozole A. 20 : 1; B,1 : 1; C,1 : 20.

---

as expected, that for Oxford staphylococcus is 1 : 20. The maximum potentiation of activity against *B. fragilis* is not as great as that against the Oxford staphylococcus. These results have two important implications. The first is that cotrimoxazole may be usefully active against most strains of *B. fragilis*, even in the inappropriate ratio present in cotrimoxazole if one takes as a guide the good results in the treatment of infection with *Neisseria gonorrhoeae*, which also responds best in vitro to mixtures containing more trimethoprim than sulphamethoxazole (Lawrence, Phillips & Nicol, 1973). However, relative resistance to either agent might, as with gonorrhoea, result in treatment failures, and the variability of the effect of cotrimoxazole on *B. fragilis* in the gut flora might also depend on relative resistance of some strains (Knothe, 1973). The second implication is that discs containing 20 parts of sulphamethoxazole and one part of trimethoprim would not be expected to detect potentiation. Fig. 4.6 shows that zones of inhibition of growth around discs containing sulphamethoxazole and trimethoprim in the proportion 20 : 1 are generally no larger than zones around discs containing sulphamethoxazole alone (Phillips & Warren, 1974).

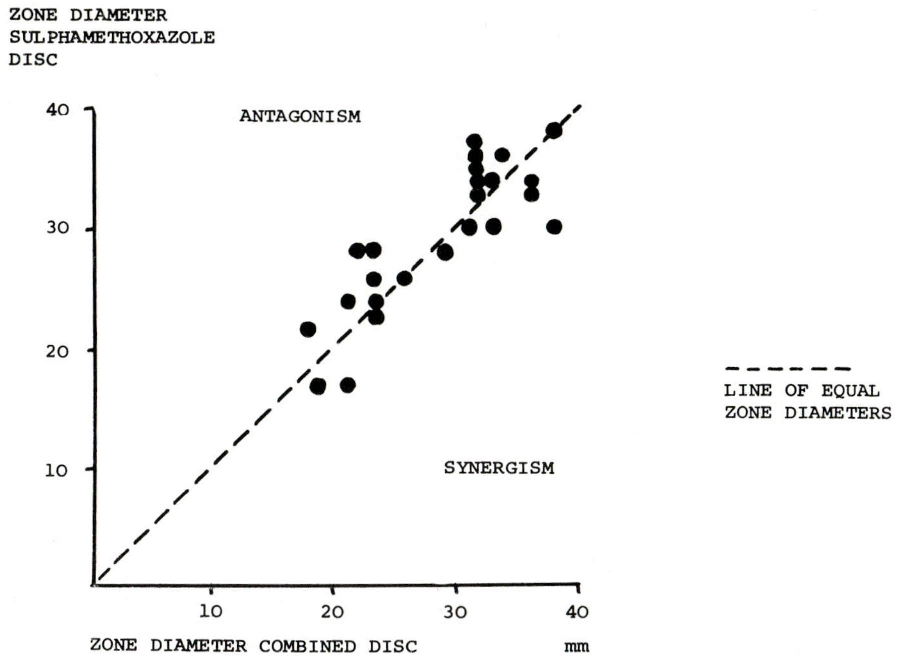

Fig. 4.6. Comparison of zones round 25 ug sulphamethoxazole discs and 23.75 ug sulphamethoxazole plus 1.25 ug trimethoprim discs for *Bacteroides fragilis*.

BACTERICIDAL ACTIVITY OF ANTIMICROBIAL AGENTS

There has been little work on the bactericidal activity of most antimicrobial agents on anaerobes, and what has been done illustrates the importance of standardization of techniques if comparable results are to be obtained. Ingham *et al.* (1968) found only ampicillin and penicillin bactericidal at concentrations a little above their MIC's, while chloramphenicol, lincomycin, and erythromycin were bactericidal at concentrations of two to four times the MIC. Nastro & Finegold (1972) reported that of clindamycin, rifampicin, vancomycin, metronidazole and tetracycline, only metronidazole was consistently bactericidal at concentrations readily achievable in serum. Rifampicin showed a 'skip-tube' phenomenon that indicated the presence of resistant mutants in most populations. In some contrast, Martin *et al.* (1972) found benzylpenicillin, chloramphenicol, and tetracycline MBC's four to eight times the MIC's and lincomycin, clindamycin and erythromycin MBC's at least 16 times the MIC's. Whelan and Hale (1973) have confirmed the bactericidal activity of metronidazole. These findings may be of importance in the choice of drugs for the treatment of infections such as endocarditis, in which a bactericidal effect is needed.

CHOICE OF ANTIBACTERIAL AGENT FOR TREATMENT

Because the isolation of anaerobes from samples obtained from patients may be delayed or at worst totally unsuccessful, treatment often has to be undertaken without direct bacteriological help. The studies reported in this review should give a useful guide in such situations. Agents that appear to have useful activity against most anaerobes include clindamycin and perhaps lincomycin, metronidazole, rifampicin, chloramphenicol, and certain penicillins, especially carbenicillin. Of these, only lincomycin and clindamycin have had their usefulness clearly demonstrated clinically (Bartlett, Sutter & Finegold, 1972; Tracy, Gordon, Moran, Love & McKenzie, 1972; Fass, Scholand, Hodges and Saslaw, 1973). Chloramphenicol is best avoided because of its potential toxicity, while rifampicin, though active against most of the important anaerobes should probably be avoided because of the risk of the emergence of resistance. Resistance is already a problem with tetracycline, but the more active minocycline might find a place. Carbenicillin has yet to be tried clinically. Yet other agents that appear promising such as fusidic acid, cotrimoxazole, and possibly spectinomycin, require further *in vitro* assessment. The only consistently bactericidal agent so far identified appears to be metronidazole.

LABORATORY USES OF ANTIBACTERIAL AGENTS

Because of the relatively predictable pattern of susceptibility of anaerobes to a variety of antibiotics, there are two laboratory uses of antibiotics that deserve mention. These are the incorporation of antibiotics in selective media for the isolation of anaerobes (Finegold, Sugihara & Sutter, 1971), and the use of disc susceptibility tests in their identification (Finegold, Harada & Miller, 1967; Finegold & Miller, 1968; Sutter & Finegold, 1971). Full details of both applications are given in a monograph by Sutter, Attebury, Rosenblatt, Bricknell & Finegold (1972). Table 4.7 gives details of antibiotics in various media and the organisms selected by them. Most are based on aminoglycosides, but vancomycin is useful in inhibiting all Gram-positive species, while rifampicin allows some *Fusobacterium* species, in particular, to grow. Table 4.8 shows characteristic disc sensitivity patterns for six antibiotics in carefully chosen amounts for the important groups of anaerobes.

Table 4.8. Antibiotic susceptibility in identification of anaerobes.

	Gram-positive cocci	Gram-negative cocci	*Bacteroides fragilis*	*melaninogenicus oralis*	*corrodens*	*Fusobacterium mortiferum*	other
colistin 10 µg	R	S	R	V	S	S	S
erythromycin 60 µg			S	S	S	R	S^r
kanamycin 1000 µg			R	R	S	S	S
penicillin 2 µ	S	S	R^s	S	S	S^r	S
rifampicin 15 µg			S	S	S	R	S
vancomycin 5 µg	S	R	R	V	R	R	R

adapted from Sutter *et al.* (1972).

R - resistant, S - sensitive, R^s - usually resistant, S^r - usually sensitive.

Table 4.7. Selective media for the isolation of anaerobes.

Antibiotics	Organism selected
kanamycin - vancomycin	*Bacteroides*
neomycin - vancomycin	*Fusobacterium*, *Veillonella*
rifampicin	*F. varium*, *F. mortiferum* certain *Eubacterium* and *Clostridium* spp.
neomycin	anaerobic cocci and *Clostridium*

adapted from Sutter *et al.* (1972).

Acknowledgements

I would like to thank Miss Christine Warren, FIMLT for performing many of the investigations at St Thomas' Hospital reported in this study, and Dr S.M. Finegold for his help in supplying details of the very substantial work of his group.

REFERENCES

Anderson, J.D. & Sykes, R.B. (1973). Characterisation of a beta-lactamase obtained from a strain of *Bacteroides fragilis* resistant to beta-lactam antibiotics. *Journal of Medical Microbiology*, 6, 201.

Bartlett, J.G., Sutter, V.L. & Finegold, S.M. (1972). Treatment of anaerobic infections with lincomycin and clindamycin. *New England Journal of Medicine*, 287, 1006.

Blake, G.C. (1964). Sensitivities of colonies and suspensions of *Actinomyces israelii* to penicillins, tetracyclines and erythromycin. *British Medical Journal*, 1, 145.

Bodner, S.J., Koenig, M.G., Treanor, L.L. & Goodman, J.S. (1972). Antibiotic susceptibility testing of bacteroides. *Antimicrobial Agents and Chemotherapy*, 2, 57.

Braude, A.I. (1973). Bacteroides in the blood. *Lancet*, 1, 377.

Bushby, S.R.M. (1973). Trimethoprim-sulfamethoxazole: in vitro microbiological aspects. *Journal of Infectious Diseases*, 128 suppl., S442.

Del Bene, V.E. & Farrar, W.E. (1973). Cephalosporinase activity in *Bacteroides fragilis*. *Antimicrobial Agents and Chemotherapy*, 3, 369.

Fass, R.J., Scholand, J.F., Hodges, G.R. & Saslaw, S. (1973). Clindamycin in the treatment of serious anaerobic infections. *Annals of Internal Medicine*, 78, 853.

Finegold, S.M., Harada, N.E. & Miller, L.G. (1967). Antibiotic susceptibility patterns as aids in classification and characterisation of gram-negative anaerobic bacilli. *Journal of Bacteriology*, 94, 1443.

Finegold, S.M. & Hewitt, W.L. (1956). Antibiotic sensitivity pattern of Bacteroides species. *Antibiotics Annual* 1955-1956, p.794. New York: Medical Encyclopedia, Inc.

Finegold, S.M., & Miller, L.G. (1968). Susceptibility to antibiotics as an aid in classification of Gram-negative anaerobic bacilli. In *The Anaerobic Bacteria* p.139-145 Ed. V. Fredette, Montreal University.

Finegold, S.M., Miller, A.B. & Posnick, D.J. (1965). Further studies on selective media for Bacteroides and other anaerobes. Enahrungsforschung Band X, Heft 2/3,,517.

Finegold, S.M., Sugihara, P.T. & Sutter, V.L. (1971). Use of selective media for isolation of anaerobes from humans. In *Isolation of Anaerobic Organisms*, p.99-108. Eds. Shapton, D.A. & Board, R.G. Academic Press, London.

Finegold, S.M. & Sutter, V.L. (1971). Susceptibility of Gram-negative anaerobic bacilli to gentamicin and other aminoglycosides. *Journal of Infectious Diseases*, 124 suppl., S 56.

Finegold, S.M. & Sutter, V.L. (1972). Antimicrobial susceptibility of anaerobic Gram-negative bacilli. In *Host Resistance to Commensal Bacteria* Ed. T. MacPhee, Churchill Livingstone, Edinburgh and London.

Garrod, L.P. (1955). Sensitivity of four species of Bacteroides to antibiotics. *British Medical Journal*, 2, 1529.

Garrod, L.P., Lambert, H.P. & O'Grady, F. (1973). *Antibiotic and Chemotherapy*, 4th Edition, Churchill Livingstone, Edinburgh and London.

Gillespie, W.A. & Guy, J. (1956). Bacteroides in intra-abdominal sepsis: their sensitivity to antibiotics. *Lancet*, 1, 1039.

Ingham, H.R., Selkon, J.B., Codd, A.A. & Hale, J.H. (1968). A study *in vitro* of the sensitivity to antibiotics of *Bacteroides fragilis*. *Journal of Clinical Pathology*, 21, 432.

Ingham, H.R., Selkon, J.B., Codd, A.A. & Hale, J.H. (1970). The effect of carbon dioxide on the sensitivity of *Bacteroides fragilis* to certain antibiotics *in vitro*. *Journal of Clinical Pathology*, 23, 254.

Keusch, G.T. & O'Connell, C.J. (1966). The susceptibility of bacteroides to the penicillins and cephalothin. *American Journal of Medical Science*, 251, 428.

Kislak, J.W. (1972). The susceptibility of *Bacteroides fragilis* to 24 antibiotics. *Journal of Infectious Diseases*, 125, 295.

Knothe, H. (1973). The effect of a combined preparation of trimethoprim and sulphamethoxazole following short-term and long-term administration on the flora of the human gut. *Chemotherapy*, 18/5, 285.

Lawrence, A., Phillips, I., & Nicol, C. (1973). Various regimens of trimethoprim-sulphamethoxazole used in the treatment of gonorrhea. *Journal of Infectious Diseases*, 128. Suppl., S673.

Leigh, D.A. (1973). Bacteroides infections. *Lancet*, 2, 1081.

Lerner, P.H. (1969). Susceptibility of *Actinomyces* species to lincomycin and its 7-halogenated analogues. *Antimicrobial Agents & Chemotherapy*, -1968, 461.

Martin, W.J., Gardner, M. & Washington, J.A. (1972). In vitro antimicrobial susceptibility of anaerobic bacteria isolated from clinical specimens. *Antimicrobial Agents & Chemotherapy*, 1, 148.

Miller, L.G. & Finegold, S.M. (1967). Antibacterial sensitivity of *Bifidobacterium (Lactobacillus bifidus)*. *Journal of Bacteriology*, 93, 125.

Mitchell, A.A.B. (1973). Incidence and isolation of Bacteroides species from clinical material and their sensitivity to antibiotics. *Journal of Clinical Pathology*, 26, 738.

Nastro, L.J. & Finegold, S.M. (1972). Bactericidal activity of five antimicrobial agents against *Bacteroides fragilis*. *Journal of Infectious Diseases*, 126, 104.

Okubadejo, O.A., Green, P.J. & Payne, D.J.H. (1973*a*). Bacteroides in the blood, *Lancet*. 1, 147.

Okubadejo, O.A., Green, P.J., & Payne, D.J.H. (1973*b*). Bacteroides infection among hospital patients. *British Medical Journal*, 2, 212.

Phillips, I. & Warren, C. (1974). Unpublished.

Pinkus, G., Veto, G. & Braude, A.I. (1968). Bacteroides penicillinase. *Journal of Bacteriology*, 96, 1437.

Rosenblatt, J.E. & Schoenknecht, F. (1972). Effects of several components of anaerobic incubation on antibiotic susceptibility test results. *Antimicrobial Agents & Chemotherapy*, 1, 433.

Sutter, V.L., Attebery, H.R., Rosenblatt, J.E., Bricknell, K.S. & Finegold, S.M. Anaerobic Bacteriology Manual, Department of Continuing Education in Health Sciences, University of California.

Sutter, V.L. & Finegold, S.M. (1971). Antibiotic disc susceptibility tests for rapid presumptive identification of Gram-negative anaerobic bacilli. *Applied Microbiology*, 21, 13.

Sutter, V.L., Kwok, Y-Y, & Finegold, S.M. (1972). Standardised antimicrobial disc susceptibility testing of anaerobic bacteria. 1. Susceptibility of *Bacteroides fragilis* to tetracycline. *Applied Microbiology*, 23, 268.

Sutter, V.L., Kwok, Y-Y. & Finegold, S.M. (1973). Susceptibility of *Bacteroides fragilis* to six antibiotics determined by standardized antimicrobial disc suscept--ibility testing. *Antimicrobial Agents & Chemotherapy*, 3, 188.

Tally, F.P., Sutter, V.L. & Finegold, S.M. (1972). Metronidazole versus anaerobes. *California Medicine*, 117, 22.
Thornton, G.F. & Cramer, J.A. (1971). Antibiotic susceptibility of *Bacteroides* species. *Antimicrobial Agents & Chemotherapy*, 1970. pp.509-513.
Tracy, O., Gordon, A.M., Moran, F., Love, W.C. & McKenzie, P. (1972). Lincomycins in the treatment of Bacteroides infections. *British Medical Journal*, 1, 280.
Whelan, J.P.F. & Hale, J.H. (1973). Bactericidal activity of metronidazole against *Bacteroides fragilis*. *Journal of Clinical Pathology*, 26, 393.

DISCUSSION

DR NAKHLA (ST ALBANS): It is known that streptomycin is inactivated under anaerobic conditions. Could this be a general property of aminoglycosides and might it explain why these organisms are generally resistant to this group of antibiotics? Have you encountered any difference between *in vitro* and *in vivo* sensitivity, in other words, the organism appearing resistant on *in vitro* testing and the patient still responding to gentamicin?
DR PHILLIPS: To take the last point first. I haven't had experience of treating these patients with gentamicin and I don't remember any that have been treated accidentally. I've a feeling that we have in fact occasionally isolated *Bacteroides* from the blood of patients who have been on gentamicin but I can't go beyond that. As far as the sensitivity is concerned, in some of our studies we did the very thing that we shouldn't have done - we used an aerobic organism, the Oxford *Staphylococcus*, as a control. There were some differences under anaerobic conditions but they were not wildly out; in other words, the antibiotic would seem to be still there and reasonably active.
DR REEVES (BRISTOL): Your definitions of sensitivity and resistance presumably must relate ultimately to the patient. Did you take into account when you said they were sensitive to fucidic acid, that only about 3 per cent of the drug in the serum is free and therefore available for antimicrobial activity?
DR PHILLIPS: These organisms have MIC's about twice, perhaps four times, those of staphylococci, and it was on these grounds that I suggested that they might be sensitive *in vivo* and *in vitro*.

PROF. GARROD: Sir, may I join you in congratulating Dr Phillips on the amount of ground he has covered and on the clarity of his delivery. I would like to ask him a question about the application of the disk test. I have tried to follow some recent American literature on this. It seems that some of the foreign media they use contain substances such as thioglycollic acid which inactivates certain antibiotics. It has always seemed to me that the rigid application of what was the Kirby-Bauer, and is now the FDA rule, about the interpretation of disk tests is utterly invalidated for these anaerobes by the fact of their extraordinarily varied rate of growth. Some of them are very slow growers and zones of inhibition are going to be correspondingly very much larger. Could I ask whether, in your studies comparing MICs with disk zone diameters, the rate of growth was apparently a factor in messing up your regression lines which I don't think you will be able to resolve statistically as you seem to think!

DR PHILLIPS: We have been using the easiest of these organisms *Bacteroides fragilis* which grows quickly in 24 hours, particularly when one has subcultured it in the laboratory. Really I must hand over to Dr Finegold who has done much more on other kinds of anaerobes to answer your question. It is of course a very considerable problem.

DR FINEGOLD: Professor Garrod's points are certainly well taken. There is considerable variation according to the rate of growth and we find that it is necessary not only to consider them separately in terms of rapid and slow growers but actually often in terms of specific species. The usefulness of the disk test therefore is considerably lessened by the fact that it is important that you speciate the organism before you can safely apply a disk procedure and this really makes it much less useful for clinical application.

DR GHONEIM (EXETER): Do you consider *Bacteroides fragilis* as a homogeneous group or does it contain a lot of different species differing in their sensitivity?

DR PHILLIPS: We've tried to subspeciate our *B. fragilis*, if this is what you mean, and they belong to the subspecies *fragilis* usually, and occasionally to the subspecies *thetaiotaomicron*. These seem to have the same susceptibility to antibiotics as far as we've gone. I haven't done any more than that and I'm not aware of a published breakdown on this one. There's no real agreement on whether they are species or subspecies - that's another problem.

5. Ecology and Metabolic Activity of Non-sporing Anaerobes

S. TABAQCHALI

Although in recent years there have been major advances in the study of intestinal bacteriology, our knowledge of the intestinal flora is still far from adequate. However, the introduction of improved anaerobic techniques has contributed greatly to a better understanding of the micro-ecology of the gastro-intestinal tract and the application of modern biochemical methods has helped to unravel some of the complex metabolic activities of these bacteria in both health and disease.

Although we now know that, in most cases, the intestinal bacterial flora appears to live in happy symbiosis with the host, there are certain conditions in which the intestinal flora has been shown to cause clinical disorders, either by the direct effect of bacterial metabolites, or by interfering with normal physiological processes, leading to malabsorption and to nutritional deficiencies.

This paper describes, briefly, the normal flora of the gastro-intestinal tract, then lists the conditions in which an abnormal bacterial flora may develop within the lumen of the small intestine, and finally discusses some of the metabolic activities of the bacteria involved and their clinical significance in man.

THE NORMAL BACTERIAL FLORA OF THE GASTRO-INTESTINAL TRACT

The bacterial flora of the intestinal tract is derived from the oropharynx, saliva and food. It is now well established that there are two types of micro-flora in the intestine, the resident flora which varies very little under constant conditions, and the transient flora, which is introduced with meals and causes a wave-like increase in the microbial population.

The fasting normal stomach in man is virtually sterile because of the presence of acid. After a meal, however, counts of up to 10^5 organisms per ml may be recovered (Drasar *et al.*, 1969). In the duodenum and upper jejunum there are Gram-positive micro-organisms, mainly streptococci and lactobacilli, in concentrations rarely exceeding 10^3-10^4 organisms per ml. Towards the ileum, however, the concentration increases and enterobacteria and *Bacteroides* may also be present. The concentrations may reach 10^5-10^7 organisms per ml in the distal ileum (Gorbach *et al.*, 1967*a*; Drasar *et al.*, 1969). In the caecum the situation changes and the composition is similar to that found in faeces. The non-sporing anaerobic bacteria form the bulk of the flora and aspirates yield up to 10^9 organisms per ml. (Gorbach *et al.*, 1967*a*). However, if the continuity of the intestine is interrupted, as in patients with ileostomy, the bacterial counts in ileal contents are much higher than those of normal ileal contents but lower than those of faeces, and this constitutes a unique micro-ecology (Gorbach *et al.*, 1967*b*; Finegold *et al.*, 1970; Vince, O'Grady & Dawson, 1973).

The total viable bacterial counts obtained from faeces using the very strict anaerobic techniques were in the region of 10^{10}-10^{11} organism per gram of faeces (Cato & Holdeman, 1969; Drasar, Shiner & McLeod, 1969*b*). The main groups of bacteria and the types commonly isolated, together with the concentrations per gram of faeces are shown in Table 5.1.

Table 5.1. Bacteria occurring in the intestine.

Major bacterial groups	Counts/g wet wt faeces	Species often isolated
Enterobacteria	10^7-10^8	*Escherichia coli* *Klebsiella aerogenes* *Proteus mirabilis*
Streptococcus	10^6-10^7	*Streptococcus viridans* *Streptococcus faecalis*
Lactobacillus	10^5-10^6	*Lactobacillus acidophilus* *Lactobacillus casei*
Clostridium	10^4	*Clostridium perfringens* *Clostridium sporogenes*
Veillonella	10^7-10^8	*Veillonella parvula* *Veillonella alcalesceus*
Bacteroides	10^{10}-10^{11}	*Bacteroides fragilis* *Bacteroides melaninogenicus*
Bifidobacterium	10^{10}-10^{11}	*Bifidobacterium adolescentis*
Eubacterium	10^{10}-10^{11}	*Eubacterium biforme*

Anaerobic bacteria constitute the majority of the faecal flora, making up approximately 98-99 per cent of the total count, *Bacteroides* and bifidobacteria being the dominant organisms. The aerobic flora is predominantly represented by enterobacteria, *Escherichia coli* being the commonest, and by *Streptococcus faecalis* and aerobic lactobacilli. Other bacterial species, such as *Staphylococcus aureus*, *Pseudomonas aeruginosa* and aerobic sporing bacilli, are also present but in low concentrations. The faecal flora is, however, complex. Of 230 isolates of faecal bacteria examined from 2 people, 46 per cent belonged to a single species, *Bacteroides fragilis*, but at least 36 other species were also represented (Moore *et al.*, 1969).

The gut flora described above represents the normal indigenous intestinal flora, which is well maintained and shows very little variation in repeated sampling from the small intestine (Gorbach *et al.*, 1967*a*) and from the faeces (Moore *et al.*, 1969). There is, however, another microbial population in the gut, which is closely associated with the layer of mucus overlying the mucosa and forms an integral part of the intestinal mucosa. This may be similar to the indigenous intraluminal flora in that region (Plaut *et al.*, 1967; Nelsen & Mata, 1970). Certain animal species, especially mice, have been shown to have a well defined layer of bacteria associated with the keratinised stratified squamous epithelium of the non-secreting portion of the stomach (Savage, 1970).

FACTORS CONTROLLING AND INFLUENCING THE BACTERIAL FLORA

The controlling mechanisms that limit the growth of enteric micro-organisms to such definable and reproducible limits are extremely complex and not fully understood.

1. *Gastric acid.* This plays a role in controlling the bacterial content of the stomach and the entry of certain bacteria into the small intestine (Van der Reis, 1925; Knott, 1927; Cregan, Dunlop & Hayward, 1953; Drasar *et al.*, 1969). However, not all patients with achlorhydria have high bacterial counts in the proximal intestine and other mechanisms appear to cope with the increased load in such patients.

2. *Intestinal peristalsis.* This is an important factor in clearing the small intestine of bacteria (Dack & Petran, 1934; Dixon, 1960; Gorbach *et al.*, 1967*a*). Disturbance of or obstruction to the intraluminal flow will lead to abnormal bacterial overgrowth.

3. *Intestinal secretions, enzymes, mucus, and bile salts.* Although antibacterial substances are present in the stomach of suckling rabbits (Smith, 1966), no such factors have been demonstrated in man. The role of intestinal enzymes is unclear and mucus may only be involved in the mechanical cleansing of the small intestine associated with intestinal motility. However, bile salts, and in particular the deconjugated forms (Floch *et al.*, 1970; Percy-Robb & Collee, 1972) have long been known to have an inhibitory effect on certain types of bacteria, and this may play a part in controlling the relative concentrations of different organisms.

4. *Immunoglobulins.* The intestine is one of the major sites of immunological activity but the role of this activity in the control of the indigenous bacterial flora is not well understood. The response to intestinal pathogens and to oral administration of vaccines derived from intestinal pathogens such as *Vibrio cholerae*, *Salmonella typhi* and *Shigella* has been shown to be primarily a local immune response of the secretory IgA type followed by a systemic response (Freter, 1969; Fubara & Freter, 1972; Dupont *et al.*, 1972). It has been suggested that secretory IgA may play a role in activating the lysis of bacteria by complement and lysozyme (Adinolfi *et al.*, 1966). Secretory IgA may prevent the adherence of bacteria to mucosal surfaces, thus preventing colonisation (Williams & Gibbons, 1972; Gibbons & Van Haute, 1971; Freter, 1969) and bacterial attachment, thus assisting in clearing the lumen (McClelland *et al.*, 1972). Patients with immunoglobulin deficiencies have higher bacterial counts in their small intestine than normal people (Hersh *et al.*, 1970; Parkin *et al.*, 1972; Brown, Hill & Richards, 1971) which illustrates the importance of intestinal immune factors in controlling the normal flora as well as in preventing the establishment of pathogens.

5. *Bacterial interactions.* These play an important role in determining both the total numbers and the relative frequency of various species in the gut. A few aspects of bacterial interactions are:

i. The growth and metabolism of the aerobic flora may promote the growth of the strictly anaerobic bacteria by reducing the intraluminal Eh to a low level (Meynell, 1963).

ii. The indigenous flora may limit or inhibit the growth and colonisation of the gut by pathogens or artificially introduced organisms (Buck & Cooke, 1969; Maier & Hentges, 1972).

iii. Cross-feeding interactions, such as the utilisation of B vitamins and amino acids produced by certain species of bacteria, and K vitamins biosynthesised by *Bacteroides fragilis* and *Veillonella* (Gibbons & Engle, 1964).

iv. The production by one organism of a bacterial metabolite that is essential for the growth of another, as, for example, the production by *Bacteroides ruminicola* of branched-chain volatile acids, such as 2-methylbutyrate, which are essential for the growth of another anaerobe, *Bacteroides succinogenes* (Bryant *et al.*, 1971).

v. Many anaerobes produce fatty acids that inhibit the growth of other bacteria (Meynell, 1963).

Bacterial interaction, is therefore, a very complex and important aspect of the control of indigenous bacteria (Donaldson, 1968).

6. *Environment and Diet.* There is some evidence that environment and diet may have an effect on the gut flora. Healthy people in India (Gorbach *et al.*, 1969) and Guatemala (Nelson & Mata, 1970) appear to have more bacteria in the small intestine than people in more temperate countries. However, the nutritional status of these individuals may play a part (Gorbach, 1971). Different diets may alter the relative numbers of various groups of bacteria. Thus, people in Britain and U.S.A. on high fat, high protein diets have more *Bacteroides* and fewer enterococci than people living largely on a vegetarian diet in Uganda or South India (Hill *et al.*, 1971). There are conflicting reports on the effect of a partially chemically defined diet on normal human faecal flora (Attebery, Sutter & Finegold, 1972).

CONDITIONS ASSOCIATED WITH BACTERIAL OVERGROWTH

The conditions under which the small intestine develops an abnormal growth of bacteria are listed in Table 5.2. There are three main types of gastro-intestinal abnormality that encourage bacterial growth. The first group includes abnormalities of the stomach, such as achlorhydria and partial gastrectomy. The second group consists of conditions causing intestinal stasis. These include strictures, whether congenital, infective or associated with Crohn's Disease (Vince *et al.*, 1972); stagnant intestinal loops following surgical enteroanastomosis; intestinal diverticulosis (multiple or single); conditions inhibiting normal intestinal motility, such as scleroderma (Kahn Jeffries & Sleisenger, 1966; Salen Goldstein & Wirts, 1966) and intestinal pseudo-obstruction (Dyer *et al.*, 1969). Tropical sprue may also be associated with abnormal bacterial growth in the ileum (Gorbach *et al.*, 1969), and partial biliary obstruction with cholangitis may act as a source of bacterial contamination of the small intestine (Scott & Kahn, 1968). The third group consists of conditions in which there is free communication between large and small bowel through a fistula, as in ileocolic fistula due to Crohn's disease or diverticulitis of the colon. Massive intestinal resection may also lead to bacterial contamination, particularly when the ileocaecal valve has been resected (Gorbach & Tabaqchali, 1969; Tabaqchali, 1970*a*).

Table 5.2. Conditions associated with an abnormal bacterial flora.

1. Abnormalities of gastric function
   Polya partial gastrectomy - Afferent loop syndrome
   Malfunctioning gastrojejunostomy
   Pernicious anaemia

2. Conditions causing stasis
   a) Surgical blind loops
      Entero-anastomosis
   b) Strictures
      Congenital
      Crohn's disease
      Tuberculous
   c) Adhesions
      X-ray irradiation
   d) Small intestinal diverticulosis
   e) Abnormal motility
      Scleroderma
      Tropical sprue
      Diabetic neuropathy
      Vagotomy
      Ganglion blocking agents
      Intestinal pseudo-obstruction
   f) Partial biliary obstruction with cholangitis

3. Free communications between large and small bowel
   Gastrocolic fistula
   Enterocolic fistula
   Massive intestinal resection

In all these conditions the bacterial flora of the small intestine differs from the normal, and the bacteria isolated are similar to those found in the faeces. Aerobic and anaerobic bacteria are present in relatively high numbers but the concentrations and the distribution depends on the type, site and extent of the causative intestinal lesion and vary from 10^5-10^9 organisms per ml of intestinal aspirate.

METABOLIC ACTIVITIES OF BACTERIA

THE METABOLISM OF BILE ACIDS

Primary bile acids, cholic acids (3α, 7α, 12α) and chenodeoxycholic acid (3α, 7α), are synthesised from cholesterol in the liver, conjugated with glycine or taurine, and secreted in the bile as conjugated bile acids. Another constituent of normal bile is the secondary bile acid deoxycholic acid (3α, 12α) which is produced by bacterial action on cholic acid. These conjugated bile acids contribute to fat digestion and absorption in the proximal small bowel and are largely reabsorbed by an active transport mechanism in the ileum which returns them to the liver via the portal vein, thus forming the enterohepatic circulation. Small amounts of unabsorbed bile acids pass into the caecum and colon where they are metabolised by bacteria.

Bacterial metabolism of bile acids is illustrated in Fig.5.1. The main reactions carried out by bacteria are:

1. *Hydrolysis:* bacteria hydrolyse the amide bond to release free bile acids from their taurine or glycine conjugates. The bacterial enzyme responsible is cholyglycine peptide hydrolase.
2. *7α Dehydroxylation:* bacteria dehydroxylate by removing the hydroxyl group at the C7 position to form the secondary bile acids. The enzymes are 7α dehydroxylases.
3. *Oxido-reduction and dehydrogenation:* bacteria oxidise and reduce the bile acid molecule to form ketoacids and α and β epimers. This can occur at the C_3, C_7 and C_{12} positions, and enzymes responsible being dehydrogenases.

These bacterial metabolites of bile acids have been identified in the faeces.

METABOLISM OF BILE ACIDS

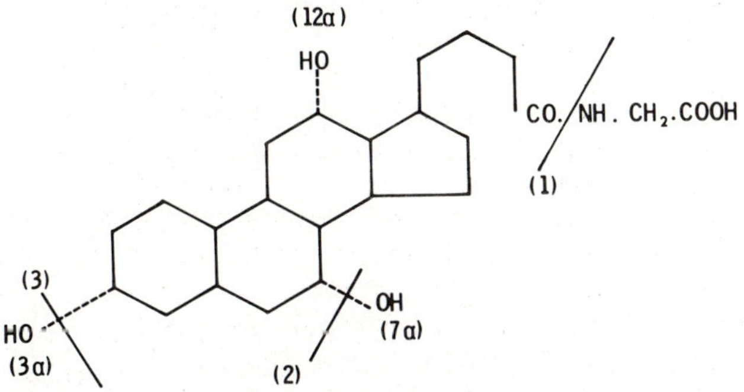

Fig.5.1. Schematic representation of glycocholic acid molecule. Sites of bacterial metabolism are denoted by reactions: (1). deconjugation, (2). 7α dehydroxylation and (3). Oxido-reduction and dehydrogenation.

Micro-organisms associated with bile acid metabolism

Various aerobic and anaerobic bacteria have been shown to be capable of carrying out these metabolic activities *in vitro* (Midvedt & Norman, 1967; Hill & Drasar, 1968). The various bacteria isolated from the faeces, with their relative counts per gram of faeces and the proportion positive for the presence of hydrolase, 7α dehydroxylase and the dehydrogenases, are shown in Table 5.3 (Aries & Hill, 1970*a*).

Table 5.3. Percentage of human intestinal bacteria capable of degrading bile salts.

Bacteria	Counts/g wet wt. faeces	Number of strains having high enzyme activity (%)				
		Hydrolase	7-Dehydroxylase	Hydroxycholanyl dehydrogenase		
				7α-	3α-	12α- (approx.)
Enterobacteria	10^7	0	2	78	39	10
Enterococci	10^6	93	11	81	29	10
Clostridia	10^4	94	34	87	47	10
Bacteroides	10^{10}	82	44	79	30	10
Bifidobacteria	10^{10}	74	40	56	21	10

Aries & Hill, 1970

A variety of micro-organisms, both aerobic and anaerobic, therefore possesses these enzymes, but, quantitatively, because of their relative high numbers, it is the anaerobic bacteria that are by far the most important.

Table 5.4 (Gorbach & Tabaqchali, 1969) shows the results of an analysis of 81 different small intestinal aspirates from patients with bacterial overgrowth in the small intestine. Free bile acids were present in 31 of these samples but only conjugated bile acids were found in the remaining 50 specimens. There was a close correlation between the presence of free bile acids and the anaerobic *Bacteroides* in these samples. Of the 31 samples in which free bile acids were found, *Bacteroides* were present in 29. In the 50 samples without free bile acids, only one contained *Bacteroides*. No similar correlation was found between the presence or absence of bile salt deconjugation and the other micro-organisms isolated.

This correlation of anaerobic bacteria and the presence of free bile acids in individual subjects is illustrated in Fig.5.2 (Tabaqchali, 1970*b*). It demonstrates the distribution of two micro-organisms only, the aerobic coliforms and the anaerobic *Bacteroides*, at different sites in the small intestine in patients with a Polya partial gastrectomy, a single duodenal diverticulum, and a stricture of the ileum, together with the presence or absence of bile salt deconjugation in each sample.

Table 5.4. Micro-organisms associated with deconjugation of bile salts in small bowel fluid.

Free Bile Acid	Samples	Isolation of Intestinal Micro-organisms in Small Bowel Fluid									
		Bacteroides		Clostridia		Bifido-bacteria		Faecal Streptococci		Enyrto-bacteria	
		Number	+ve	Number	+ve	Number	+ve	Number	+ve	Number	+ve
Present	31	29	(94%)	15	(49%)	13	(42%)	29	(94%)	31	(100%)
Absent	50	1	(2%)	9	(18%)	13	(26%)	30	(60%)	48	(96%)
Total	81	30		24		26		59		79	

Gorbach & Tabaqchali, 1969

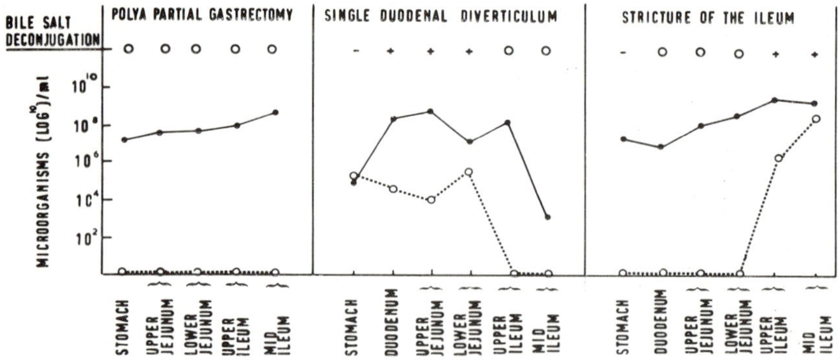

Fig.5.2. The concentration and distribution of two micro-organisms, the aerobic coliforms ●—● and the anaerobic *Bacteroides* ○······○ at different sites of the gastro-intestinal tract of 3 patients (Polya partial gastrectomy, single duodenal diverticulum, and stricture of the ileum) together with the presence or absence of bile salt deconjugation in each sample.

In patients with an uncomplicated Polya partial gastrectomy, where there is no stasis, coliforms may be present throughout the small intestine, but no *Bacteroides* are isolated and no free bile acids detected. In the other two patients with a single duodenal diverticulum and with an ileal stricture, where the lesions are limited, there were coliforms throughout the small intestine, but *Bacteroides* were only present in the areas of stasis, proximally in the first patient and distally in the patient with the ileal lesion. At the same time free bile acids were present in these samples only.

These results, therefore, suggest that although *in vitro* a variety of micro-organisms have been shown to be capable of deconjugating and dehydroxylating the bile acids, *in vivo* this is usually associated with the presence of *Bacteroides*, and that the presence of *Bacteroides* may serve as a marker for the conditions under which enzymatic activity for bile salt degradation occurs.

In patients with bacterial overgrowth in the small intestine where bile salt degradation occurs, the bile acid pattern in the intestinal lumen will be deranged. In Fig. 5.3 the bile acid concentration in fasting jejunal fluid from 5 control subjects (hatched columns) and in 7 patients with the stagnant loop syndrome (open columns) is compared. The mean and range of the bile acids are expressed as

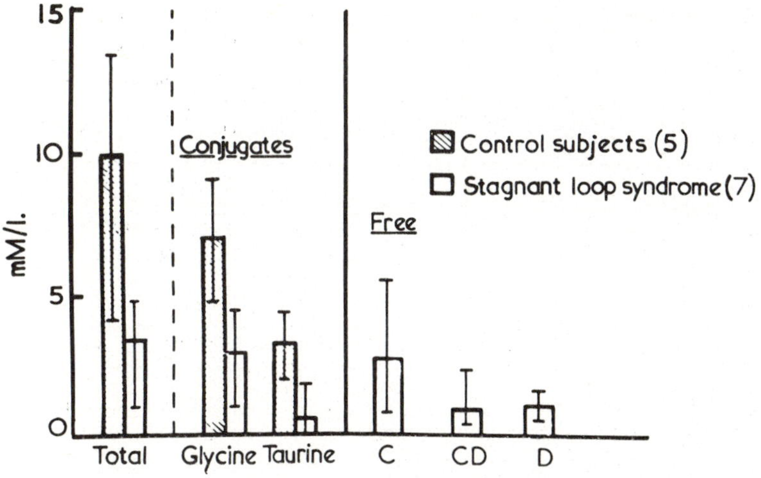

Fig. 5.3. The concentrations of bile acids in the fasting upper jejunal fluid (mean and range in mmoles/litre) from 5 control subjects (hatched areas), and 7 patients with the stagnant loop syndrome (open areas). Total concentration of conjugated bile acids and their subdivision into glycine and taurine conjugates are represented. The free bile acids are represented by C=cholic acid, CD=chenodeoxycholic acid and D=deoxycholic acid.

---

mmoles per litre of intestinal aspirate. The conjugated bile acids are presented as a total which is sub-divided into glycine and taurine conjugates. The free bile acids detected were cholic, chenodeoxycholic and deoxycholic acids. In the jejunum of control subjects only conjugated bile acids were detected. In the patients with bacterial overgrowth (stagnant loop syndrome) the pattern was different. Firstly, there were large amounts of free bile acids present (cholic, chenodeoxycholic and deoxycholic acids). Secondly, the taurine conjugates were much more reduced than the glycine oonjugates, and thirdly, as a result of bacterial formation of free bile acids, the level of the conjugated bile salts fell each being less than 5 mmoles per litre.

Consequences of bacterial degradation of bile acids

1. *Steatorrhoea.* The small intestine normally contains conjugated bile salts in sufficient concentrations to promote the dispersion and absorption of lipids through the formation of mixed micelles. Under abnormal conditions, where there is colonisation of the small intestine, bile salt deconjugation and dehydroxylation may occur and thus lead to a reduction in the level of the conjugated bile salts sufficient to impair micelle formation, thereby causing malabsorption of fat (Kim *et al.*, 1966, Tabaqchali, Hatziaonnou

& Booth, 1968). On the other hand, the free bile acids at the pH of intestinal contents, appear to be either absorbed by non-ionic diffusion or precipitated. However, the mere presence of abnormal bacterial overgrowth, or the presence of free bile acids in the small intestine, does not necessarily lead to steatorrhoea. Limited lesions of the jejunum or ileum may not be associated with steatorrhoea if there is insufficient degradation of bile salts (Gorbach & Tabaqchali, 1969). This subject has been reviewed in detail elsewhere (Tabaqchali, 1970*b*).

2. *Diarrhoea.* The mechanism of the diarrhoea associated with bacterial overgrowth in the small intestine and after intestinal resection is not well understood. The absorption of free bile acids has been shown to lead to a disturbance of water and electrolyte absorption in patients (Mekhjian, Phillips & Hofman, 1971) and in experimental animals (Harries & Sladen, 1972).

3. *Monosaccharide malabsorption.* Glucose transport by jejunal mucosa was impaired by the presence of deconjugated bile acids *in vitro* (Pope, Parkinson & Olson, 1966) and in experimental animals *in vivo* (Tabaqchali & Booth, 1966; Harries & Sladen, 1972). Gracey, Burke & Anderson, (1969), and Gracey, Burke & Oslin, (1971), demonstrated reversible inhibition of active intestinal sugar transport by deconjugated bile salts and suggested that this may explain the temporary monosaccharide malabsorption of infancy.

Detection of bile acid deconjugation.

The presence of bile acid deconjugation in the small intestine can be detected without intestinal intubation by two methods. These are the demonstration of free bile acids in the serum since the serum bile acids pattern reflects the changes that occur in the small intestine (Lewis *et al.*, 1969; Panvelliwella *et al.*, 1969) and the Breath Test which consists of the measurement of $^{14}CO_2$ in expired air following an oral dose of ^{14}CO glycocholic acid. Bacteria deconjugate the bile acids and further metabolise the glycine, forming $^{14}CO_2$ (Fromm & Hofmann, 1971). However, these tests are not specific for bacterial overgrowth in the small intestine since they are also positive in conditions with bile salt malabsorption and in biliary tract infections.

METABOLISM OF AMINO ACIDS

Amino acids are metabolised by gut bacteria by deamination, decarboxylation, and various other reactions. They are utilised both as carbon sources for energy production and as nitrogen sources for various synthetic activities.

The resulting end-metabolites may be absorbed from the human intestine and metabolised or excreted. Thus, bacterial ammonia may supply the nitrogen required for the synthesis of amino acids from their α-keto acid analogues (Richards *et al.*, 1971). On the other hand, many of the absorbed bacterial metabolites are rapidly excreted in the urine. These urinary metabolites will be present in larger amounts where there is bacterial overgrowth in the small intestine. Thus, rats with surgically-created blind loops of small intestine excrete more indoxyl sulphate (Indican), indole-3-acetic acid, volatile phenols, hippuric acid, and the amines piperidine and pyrolidine, as the bacterial metabolites of various amino acids, than do control rats. (Table 5.5), (Miller *et al.*, 1971). The administration of oral antibiotics causes a marked reduction in the excretion of these metabolites and at the same time a reduction in the anaerobic bacteria in the small bowel (Miller *et al.*, 1971).

Indicanuria.

The production of indole is due to a specific bacterial degradation of the amino acid tryptophan. The bacteria involved are *Esch. coli* and *Bacteroides* which possess the enzyme tryptophanase necessary for this reaction. Indoles are equally well absorbed from small and large bowel, conjugated in the liver to indoxyl sulphate and excreted in the urine. The normal excretion of indican is less than 100mg per 24 hrs. This level may be raised in patients with bacterial overgrowth and other malabsorption disorders (Tabaqchali *et al.*, 1966; Greenberger, Saegh & Ruppert, 1968; Hamilton *et al.*, 1970), but these findings should be interpreted with special regard to factors that may affect indole production by bacteria, such as diet, presence of substrate, type of micro-organism cononising the bowel, pH of the intestinal contents (Neale *et al.*, 1972) and the presence of pancreatic disease (Fordtran, Scroggie & Polter, 1964).

Severe protein malnutrition.

Disturbance of protein metabolism by bacterial overgrowth, if associated with malabsorption, may be severe enough to cause protein malnutrition, a clinical picture resembling kwashiorkor (Krickler & Schrire, 1958; Neale *et al.*, 1967). In some patients the amount of urinary indican may represent up to 70 per cent of the dietary intake of L-tryptophan. Other essential amino-acids are also degraded, thus leading to nutritional deficiency in the host. Albumin and

Table 5.5. Bacterial metabolites of amino acids excreted in the urine of control rats and blind loop rats before and after antibiotic therapy.

Amino acid precursor	Substance excreted	Control rats (n = 10)	Blind loop rats (n = 19)			
			No antibiotic	p	On Neomycin	On Lincomycin
Tryptophan	Indican	3.10 ± 2.10	7.00 ± 3.40	<0.01	3.84 ± 0.40	3.94 ± 1.41
	Indole-3-acetic acid	0.33 ± 0.08	0.43 ± 0.16	NS	0.14 ± 0.06	0.12 ± 0.03
Tyrosine	Volatile phenols	0.75 ± 0.30	2.50 ± 1.10	<0.05	0.26 ± 0.23	0.26 ± 0.18
Phenylalanine	Hippuric acid	7.80 ± 2.50	12.50 ± 3.80	<0.01	2.46 ± 1.00	2.37 ± 0.76

NS = not significant

fibrinogen synthesis may be reduced but urea synthesis is increased, suggesting that bacteria are utilising nitrogen to form ammonia which is then diverted through the urea cycle (Jones *et al.*, 1968).

Intestinal infantilism.

In children, the long-continued caloric undernutrition due to bacterial overgrowth may lead to failure of growth and sexual development, leading to a clinical picture of infantilism (Bayes & Hamilton, 1969; Neale, 1968).

Metabolism of tyrosine.

Various early workers have demonstrated that tyrosine is metabolised by the bacterial flora of the gut, to yield a number of phenolic compounds and that there was a direct relationship between dietary tyrosine and the amount of phenols in the urine. Recently, the production of phenols has been studied extensively by Bakke (1969*a,b*). He has repeated and confirmed earlier work and has also shown that treatment with neomycin reduces the amount of phenols to almost zero in the rat. Furthermore, *in vitro* incubations, under anaerobic conditions, of various urinary metabolites of tyrosine with caecal bacteria from rats, were carried out in order to establish the pathways of tyrosine metabolism (Bakke, 1969*a*).

Phenols in the urine have been suggested to be possible co-carcinogens and convulsants, and implicated in portosystemic encephalopathy associated with liver disease.

Metabolism of 3,4,-dihydroxyphenylalanine (dopa).

The recent resurgence of interest in the metabolism of L-dopa stems from its new therapeutic use in the treatment of Parkinsonism. Biotransformation of L-dopa by intestinal micro-organisms was suggested by Sandler *et al.*, (1969) who found that the quantity of m-hydroxyphenylacetic acid in the urine of patients with Parkinsonism treated with large doses of L-dopa decreased significantly when neomycin was given orally. O'Gorman *et al* (1970) demonstrated that in man the metabolites of L-dopa include 3,4-dihydroxyphenylacetic acid, m-hydroxyphenylacetic acid, homovanillic acid, and homoisovanillic acid together with the other simple phenols. Bakke (1971) demonstrated these metabolites in the rat and found that p-cresol may also be formed. O-methylation of simple phenols by rat intestinal bacteria was demonstrated and suggested as the pathway for the formation of 4-methylguaiacol from 4-methylcatechol (Bakke, 1970).

Metabolism of urea.

The deamination of urea to CO_2 and ammonia is carried out by many of the normal bacteria of the gut including the anaerobic *B. fragilis, Cl. perfringens, Bifidobacterium* sp.and *Eubacterium*. These produce a urease and hydrolyse the urea that diffuses into the gut with the production of ammonia which is reabsorbed. This may lead to recycling of nitrogen in some circumstances (Brown, Hill & Richards, 1971). However, although the ammonia normally produced by colonic bacteria can be metabolised by the liver in normal persons and causes no problem, if the liver is diseased or by-passed by the portal blood, the ammonia content of the blood may be elevated and lead to an encephalopathy (Fisher & Falcon, 1957).

METABOLISM OF VITAMIN B_{12}

Another aspect of bacterial activity that may lead to nutritional deficiency in patients with bacterial overgrowth, is the uptake of vitamin B_{12} by the bacteria, making it unavailable to the host. This may lead to vitamin B_{12} malabsorption and thus to vitamin B_{12} deficiency and megaloblastic anaemia. Treatment with oral broad spectrum antibiotics improves the vitamin B_{12} absorption, if the ileum is intact and has not been short-circuited (Mollin, Booth & Baker, 1957). The mechanism by which bacteria interfere with vitamin B_{12} absorption is still not fully understood. Three possible mechanisms have been investigated.

1. *Competitive uptake of vitamin B_{12} by micro-organisms.* Although it is well known that bacteria can take up free vitamin B_{12} *in vitro*, the addition of intrinsic factor (IF) appears to protect against its uptake *in vitro* but not *in vivo*.

The results of *in vitro* uptake of unbound and IF-bound vitamin B_{12} by different micro-organisms isolated from intestinal aspirates of patients with vitamin B_{12} malabsorption are shown in Fig.5.4. Although IF appears to protect against the uptake of vitamin B_{12} by the bacteria, this protection is not complete for certain strains of enterobacteria and in particular for certain strains of *Bacteroides*. Other factors may also be involved, such as the length of the incubation period (Donaldson, 1962), and increasing the nutrient available to the micro-organisms may alter the amount of uptake (Dellipiani, Samson & Girdwood, 1967). *In vitro*, certain bacteria have an avidity for B_{12} approaching that of human IF. IF-bound vitamin B_{12} may interchange with B_{12} on the surface of the

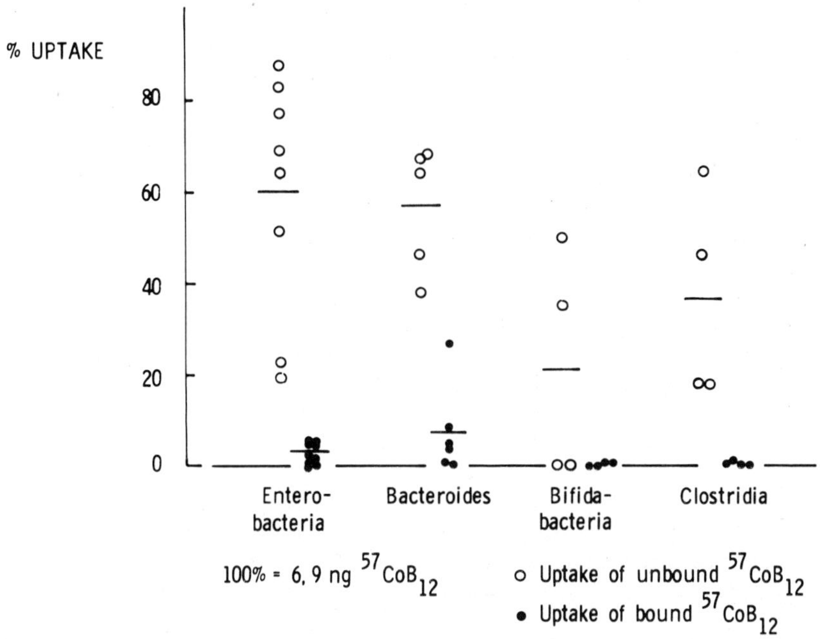

Fig. 5.4. The *in vitro* incubation of different types of micro-organisms isolated from the intestinal aspirates of patients with vitamin B_{12} malabsorption, with ^{57}Co-vitamin B_{12} unbound (open circles) and IF-bound (closed circles). Each dot represents the mean of three incubations for a different organism.

bacteria but not with the B_{12} within the interior of the cell (Granella, Broitmann & Zamcheck, 1971, 1972).

Studies in man support the hypothesis that bacteria can take up bound vitamin B_{12} *in vivo*. Patients with bacterial overgrowth (the stagnant loop syndrome) in whom there was vitamin B_{12} malabsorption, patients with bacterial overgrowth with normal vitamin B_{12} absorption, and control subjects were intubated to the ileum. After an overnight fast a test meal was administered containing 1 μg ^{57}Co vitamin B_{12} bound to human IF. Ileal aspirates were obtained at different times for 5 hours and centrifuged. The radioactivity in the deposit averaged 46-74 per cent in patients with the stagnant loop syndrome in whom there was vitamin B_{12} malabsorption, whereas it was less than 10 per cent in the other two groups (Fig.5.5). Treatment of one of these patients with antibiotics led to a reduction in the percentages of radioactivity in the deposit. At the same time there was a decrease in the anaerobic flora colonizing the small intestine, and the B_{12} absorption became normal (Fig.5.6) (Tabaqchali, 1970*b*; Schjönsby *et al.*, 1973).

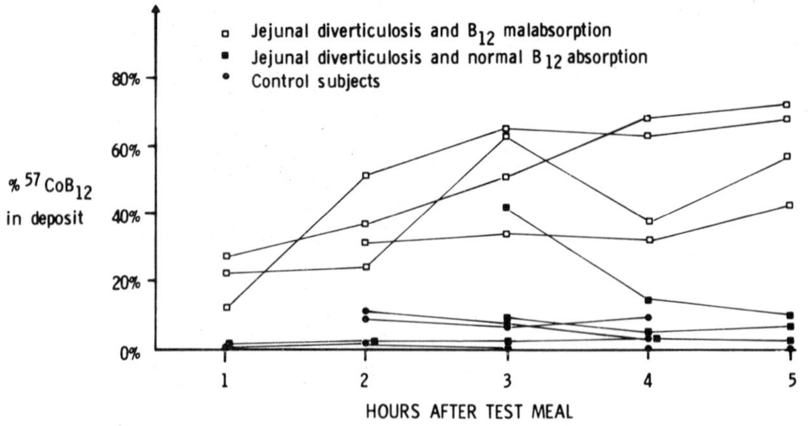

Fig. 5.5. Percentage $^{57}CoB_{12}$ in centrifuged deposits of upper ileal aspirates after a test meal containing 1 µg human-gastric-juice-bound $^{57}CoB_{12}$ in patients with jejunal diverticulosis and in control subjects.

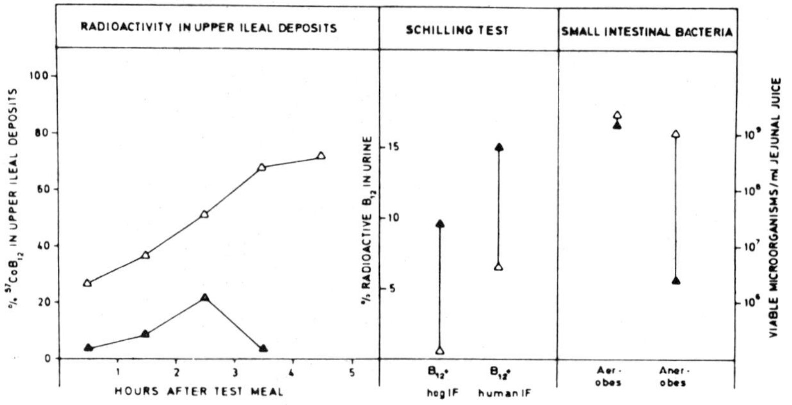

Fig. 5.6. Effect of treatment with oral antibiotics on the percentage $^{57}CoB_{12}$ in upper ileal deposits, on vitamin B_{12} absorption as assessed by Schilling test and on the microbial concentrations in upper ileal fluid in a patient with jejunal diverticulosis. Open triangles show results before, and closed triangles after antibiotic therapy.

---

2. *Destruction of intrinsic factor.* The possibility that bacteria may destroy IF activity has also been investigated. IF binding capacity was decreased (by up to 25 per cent) after incubation of human gastric juice with micro-organisms isolated from the intestine (*E. coli*, *Klebsiella* and *Bacteroides*) for 6-24 hours, but the ability of the bacterial

supernatant to bind vitamin B_{12} and its subsequent uptake by guineapig brush borders was not impaired (Schjönsby & Tabaqchali, 1971*b*).

3. *'Toxic factors'*. The theory that bacteria or their metabolites may damage ileal enterocyte receptors and thus cause vitamin B_{12} malabsorption, i.e. the 'toxic factor' theory, has also been investigated. Donaldson (1962) showed normal vitamin B_{12} absorption by the isolated ileum of rats with experimental blind loops, and Schjönsby and Tabaqchali (1971*a*) isolated mid-intestinal brush borders from control rats, and from two groups of rats with experimental blind loops (a group with normal vitamin B_{12} absorption and group with abnormal vitamin B_{12} absorption) and measured the uptake of free and rat IF-bound vitamin B_{12} by the brush borders isolated from these three groups. The results showed that the addition of IF enhances the uptake of vitamin B_{12} by the brush borders but there was no significant difference between the three groups. It appears, therefore, that bacteria do not interfere with the uptake of vitamin B_{12} by the brush borders, but whether there is interference with vitamin B_{12} transport within the enterocyte has not yet been determined.

In summary, the evidence so far suggests that bacteria interfere with vitamin B_{12} absorption in patients with bacterial overgrowth by competitive uptake of the vitamin within the lumen of the small intestine even when it is bound to IF, and that there may be minimal destruction of IF activity. The bacteria predominantly involved appear to be *Bacteroides*.

METABOLISM OF OTHER B VITAMINS

It is well known that various bacteria can synthesise B vitamins and also utilise them, but there is no clear evidence whether this may have clinical significance in man. Signs of vitamin B deficiency states are not uncommon in malnourished patients but these have not been related to the intestinal bacterial flora. Tabaqchali & Pallis (1970) described one patient with jejunal diverticulosis who developed an encephalopathic syndrome with EEG changes, resembling acute nicotinic acid deficiency states, and the patient responded rapidly to treatment with a single injection of nicotinic acid (100 mg intravenously). Studies on *in vitro* incubation of faecal anaerobic micro-organisms showed that the concentration of nicotinic acid in the culture medium was decreased by 5 of 14 different bacteria and in none was it increased, whereas the concentrations of other B vitamins was variable(Gall, 1970).

METABOLISM OF FOLIC ACID

In contrast to the bacterial metabolism of citamin B_{12} making it unavailable to the host and thus leading to nutritional deficiency, bacterial synthesis of folic acid in the gut may contribute to the host's nutrition. Folic acid deficiency rarely occurs in patients with bacterial overgrowth (Barrett & Holt, 1966; Wakisaka, 1968); serum folate levels may be either normal or occasionally raised. High serum folate levels have been recorded in patients with bacterial overgrowth, but these were not due to vitamin B_{12} deficiency per se, since treatment with vitamin B_{12} did not reduce the serum level, and the urinary excretion of folate was also markedly raised (Hoffbrand, Tabaqchali & Mollin, 1966). Bacteria in the lumen of the intestine may synthesize folic acid, which is then absorbed, increasing the serum levels in patients with distal stagnant loop syndrome, where the bacterial overgrowth is mainly in the ileum (Hoffbrand *et al.*, 1971). This may be due to the preferential absorption of folic acid from the jejunum as compared to the ileum (Hepner *et al.*, 1968). Strains of bacteria isolated from the jejunal aspirates of patients with bacterial overgrowth in whom the serum folate was high, were tested for the production and utilization of (*L.casei* active) folate. The findings are shown in Table 5.6. The principal organisms tested were *Esch. coli*, *Bacteroides* sp. and lactobacilli. Both *Esch. coli* and *Bacteroides* appear to synthesise folic acid whereas the lactobacilli utilize it from the media. The folate compound circulating in the serum of patients with stagnant loop syndrome is the same as in normal subjects, 5-methyl tetrahydrofolate (Hoffbrand *et al.*, 1971). These findings have been confirmed by recent work in dogs with experimental blind loops (Berstein *et al.*, 1972).

THE METABOLISM OF DRUGS

The metabolism of drugs and other foreign compounds, such as food additives, by intestinal bacteria has only recently received serious consideration (Williams, 1972; Scheline, 1968, 1973). Bacteria play an important role in determining the fate of these compounds and it is important to realise that bacterial metabolism may result in an increase in biological activity or toxicity. The gut microflora is capable of carrying out numerous reactions such as hydrolysis of glucuronides and glucosides, dehydroxylation, decarboxylation, dealkylation, aromatization, and various others (Scheline, 1968).

Table 5.6. Bacterial synthesis and utilisation of (*L.casei* active) folate.

Organism	Incubation			Free Folate Activity (mμg/ml)	
	Medium	Time (hr)	Atmosphere	Before	After
Esch. coli (1)	Serum	24	Aerobic	0.2	7.5
Esch. coli (2)	Serum	24	Aerobic	0.8	3.2
Bacteroides (1)	Serum	48	Anaerobic	1.1	5.0
Bacteroides (2)	Serum	48	Anaerobic	1.1	20.0
lactobacilli	*Lactobacillus* broth	24	Aerobic	2.4	1.4
	Lactobacillus broth + pteroylglutamic acid	24	Anaerobic	21.8	1.5

Some of the consequences of the biotransformation of drugs are illustrated below.

Interference with the enterohepatic circulation. Numerous drugs are conjugated in the liver with glucuronide and excreted in the bile. If bacteria hydrolyse these compounds liberating the free compound, it can be reabsorbed. This will lead to prolongation of the time the compound is in the body and an interference with the biliary excretion of drugs. Stilboestrol is an example.

Production of active drugs. The production of active drugs from inactive precursors by gut bacteria is a field requiring further investigation. The liberation of the antibacterial agent sulphanilamide from Prontosil and Neoprontosil by the splitting off the azo-links is an example (Gingell, Bridges & Williams, 1971). The release of sulphapyridine from salazopyrine may represent another bacterial metabolic activity.

The production of toxic metabolites. A well known example of this is the hydrolysis of the glycoside amygdalin by the gut flora to mandelonitrile which is unstable and yields cyanide. Amygdalin is more toxic to mice when given by mouth than when injected intraperitoneally. Another example is the hydrolysis of Lanotoside to yield digitoxin by gut bacteria.

Carcinogenesis. Epidemiological studies have shown that there is a good correlation between the incidence of colon cancer and the amount of dietary fat. There is also a good correlation between the recorded colon cancer incidence in several countries and the faecal concentration of bacterially degraded acids and neutral steroids (Hill *et al.*, 1971). Studies on the faecal flora from these different parts show a difference in that the *Bacteroides* were present in lower numbers and the enterococci in higher numbers in faeces from countries with a low cancer incidence. It has been postulated that gut bacteria metabolise bile acids to produce carcinogens, and that the metabolites produced by the bacteria are polycyclic aromatic compounds, by the dehydrogenation of the steroid nucleus. *Clostridium paraputrificum* isolated from human faeces has been shown to have this ability (Aries, Goddard & Hill, 1971; Goddard & Hill, 1972).

Breast tumours can be induced and stimulated by oestrogens (King, 1971). Therefore, the involvement of bacteria in the aetiology of cancer of the breast was suggested when it was demonstrated that intestinal bacteria could produce oestradiol, oestrone and 17-methoxyoestradiol from androstenedione (Aries *et al.*, 1971). Cholestenone is present in the faeces and is a bacterial metabolite of cholesterol. Furthermore, gut bacteria can produce 3 oxo-chol-4-en-24-oic acid from bile acids (Aries & Hill, 1970*b*).

The metabolism of nitrite and nitrate by intestinal bacteria to form nitrosamines has been suggested as the possible mechanism of carcinogenesis in the stomach of achlorhydric patients, in whom there is bacterial overgrowth and a higher incidence of cancer of the stomach (Hawksworth & Hill, 1971).

VOLATILE FATTY ACID PRODUCTION

The production of volatile fatty acids by the anaerobic metabolism of carbohydrate, fatty acids and amino acids, has been extensively studied in ruminant nutrition (Bryant, 1970). The production of volatile fatty acids in the caecum may contribute to as much as 20 per cent in the rabbit (McBee, 1970) and to 5 per cent in the rat (Yang, Manoharan & Mickelsen, 1970) of the animal's energy requirements. Moore, Cato & Holdeman (1969) have studied the production of volatile fatty acids by human colonic anaerobic bacteria and have used these products extensively in taxonomy (see p.21).

Since the production of volatile fatty acids is a primary metabolic activity of anaerobic bacteria, measurements of these acids in the intestinal content may prove a good diagnostic marker in patients with bacterial overgrowth in the small intestine. Using gas liquid chromatography, concentrations of volatile fatty acids were estimated in intestinal aspirates in normal subjects, in control patients with malabsorption without evidence of bacterial overgrowth, and in patients with the stagnant loop syndrome (Chernov, Doe & Gompertz, 1972).

The levels of acetate and propionate were raised in all patients with the stagnant loop syndrome as compared to the other two groups, whereas butyrate was less markedly changed. Acetate accounted for 85-95 per cent of the volatile fatty acids present. The concentrations of these volatile fatty acids in three patients with the stagnant loop syndrome fell to normal levels after treatment with antibiotics. Measurements of volatile fatty acids in the faeces have also been used as a monitor of antibiotic sterilisation of the gut in patients with leukaemia who are nursed in germfree isolators (Gompertz *et al.*, 1973). In man, the fate and metabolism of these volatile fatty acids and whether they contribute to energy requirements is not known.

REFERENCES

Adinolfi, M ., Glynn, A.A., Lindsay, M. & Milne, C.M. (1966). Serological property of A Antibodies to *Escherichia coli* present in human colustrum. *Immunology*, 10, 517.

Aries, V. & Hill, M.J. (1970*a*). The formation of unsaturated bile acids by intestinal bacteria. *Biochimica Biophysica Acta*, 119, 37P.

Aries, V. & Hill, M.J. (1970*b*). Degradation of steroids by intestinal bacteria. *Biochimica Biophysica Acta*, 202, 535.

Aries, V., Goddard, P. & Hill, M.J. (1971). Degradation of steroids by intestinal bacteria. *Biochimica Biophysica Acta*, 248, 482.

Attebery, H.R., Sutter, V.L. & Finegold, S.M. (1972). Effect of a partially chemically defined diet on normal human faecal flora. *American Journal of Clinical Nutrition*, 25, 1391.

Bakke, O.M. (1969*a*). Urinary simple phenols in rats fed purified and non-purified diets. *Journal of Nutrition*, 98, 209.

Bakke, O.M. (1969*b*). Studies on the degradation of tyrosine by rat caecal contents. *Scandinavian Journal of Gastro-enterology*, 4, 603.

Bakke, O.M. (1970). O-Methylation of simple phenols in the rat. *Acta pharmacologica et toxicologica*, 28, 28.

Bakke, O.M. (1971). Minor urinary metabolites of L-dopa in the rat. *Acta pharmacologica et toxicologica*, 29, 106.

Barrett, C.R., Jr., & Holt, P.R. (1966). Postgastrectomy Blind loop syndrome. Megaloblastic anaemia secondary to malabsorption of folic acid. *American Journal of Medicine*, 41, 629.

Bayes, B.J., & Hamilton, J.R. (1969). Blind loop syndrome in children. Malabsorption secondary to intestinal stasis. *Archives of Diseases of Childhood*, 44, 76.

Bernstein, L.H., Gutstein, S., Efron, G. & Wager, G. (1972). Experimental production of elevated serum folate in dogs with intestinal blind loops. Relationship of serum levels to location of the blind loop. *Gastroenterology*, 63, 815.

Brown, C.L., Hill, M.J. & Richards, P. (1971). Bacterial ureases in uraemic men. *Lancet*, 2, 406.
Bryant, M.P. (1970). Normal flora - rumen bacteria. *American Journal of Clinical Nutrition*, 23, 1440.
Bryant, M.P., Tzeng, S.F., Robinson, I.M. & Joyner, A.E.,Jr. (1971). Nutritional requirements of methanogenic bacteria. In: Anaerobic Biological Treatment Processes, Advances in Chemistry Series 105, ed. F.G.Pohland, Washington, D.C.
Buck, A.C. & Cooke, E.M. (1969). The fate of ingested *Pseudomonas aeroginosa* in normal persons. *Journal of Medical Microbiology*, 2, 52.
Chernov, A.J., Doe, W.R. & Gompertz, D. (1972). Intra-jejunal volatile fatty acids in the stagnant loop syndrome. *Gut*, 13, 103.
Cregan, J., Dunlop, E.E. & Hayward, N.J. (1953). The bacterial content of the human small intestine in disease of the stomach. *British Medical Journal*, 2, 1248.
Dack, G.M. & Petran, E. (1934). Bacterial activity in different levels of the intestine and in isolated segments of small and large bowel in monkeys and dogs. *Journal of Infectious Diseases*, 54, 204.
Dixon, J.M.S. (1960). The fate of bacteria in the small intestine. *Journal of Pathology and Bacteriology*, 79, 131.
Dellipiani, A.W., Samson, R. & Girdwood, R.H. (1967). Factors influencing the uptake of 58 Cobalt-labelled vitamin B_{12} by *Escherichia coli*. (Abstr.) *Gut*, 8 629.
Donaldson, R.N.,Jr. (1962). Malabsorption of Co^{60} labelled cyanocobalamin in rats with intestinal diverticula. I. Evaluation of possible mechanisms. *Gastroenterology*, 43, 271.
Donaldson, R.M.,Jr. (1968). Role of indigenous enteric bacteria in intestinal function and disease. In: Code, C.F., & Heidel, W., Eds, Handbook of Physiology. Volume 5. Washington, American Physiological Society, p.2807.
Drasar, B.S. (1967). Cultivation of anaerobic intestinal bacteria. *Journal of Pathology and Bacteriology*, 94, 417.
Drasar, B.S., Shiner, M. & McLeod, G.M. (1969). Studies on intestinal flora. I. Effects of diet, age and periodic sampling on numbers of faecal micro-organisms in man. *Gastroenterology*, 53, 845.
DuPont, H.L., Hornik, R.B., Snyder, M.J., Libonati, J.P., Formal, S.B. & Gangarosa, E.J. (1972). Immunity in shigellosis. I. Response of man to attenuated strains of shigella. *Journal of Infectious Diseases*, 125, 5.
Dyer, N.H., Dawson, A.M., Smith, B.F. & Todd, I.P. (1969). Obstruction of bowel due to lesion in the myentric plexus. *British Medical Journal*, 1,686.

Finegold, S.M., Sutter, V.L., Boyle, J.D. & Shimda, K. (1970). The normal flora of ileostomy and transverse colostomy effluents. *Journal of Infectious Diseases*, 122, 376.
Fisher, C.J. & Faloon, W.W. (1957). Blood ammonia levels in hepatic cirrhosis. *New England Journal of Medicine*, 256, 1030.
Floch, N.H., Binder, H.J., Filburn, B. & Gershengoren, W. (1972). The effect of bile acids on intestinal microflora. *American Journal of Clinical Nutrition*, 25, 1418.
Fordtran, J.S., Scroggie, J.B. & Polter, D.E. (1964). Colonic absorption of typtophan metabolites in man. *Journal of Laboratory and Clinical Medicine*, 64, 125.
Freter, R. (1969). Studies of the mechanism of action of intestinal antibody in experimental cholera. *Texas Reports on Biology and Medicine*, 27, 299.
Fromm, H. & Hofmann, A.F. (1971). Breath test for altered bile acid metabolism. *Lancet*, 2, 621.
Fubara, E.S. & Freter, R. (1972). Source and protective function of coproantibodies in intestinal disease. *American Journal of Clinical Nutrition*, 25, 1457.
Gall, L.S. (1970). Normal faecal flora of man. *American Journal of Clinical Nutrition*, 23, 1457.
Gianella, R.A., Broitmann, S.A. & Zamcheck, N. (1972). Competition between bacteria and intrinsic factor for vitamin B_{12}: implications for vitamin B_{12} malabsorption in intestinal bacterial overgrowth. *Gastroenterology*, 62, 255.
Gianella, R.A., Broitmann, S.A. & Zamcheck, N. (1971). Vitamin B_{12} uptake by intestinal micro-organisms. Mechanisms and relevance to syndromes of intestinal bacterial overgrowth. *Journal of Clinical Investigation*, 50, 1100.
Gibbons, R.J. & Engle, L.P. (1964). Vitamin K compounds in bacteria that are obligate anaerobes. *Science*, 146, 1307.
Gibbons, R.J. & Van Houte, J. (1971). Selective bacterial adherence to oral epithelial surfaces and its role as an ecological determinant. *Infection and Immunity*, 3, 573.
Gingell, R., Bridges, J.W. & Williams, R.T. (1971). The role of gut flora in the metabolism of prontosil and neopronosil in the rat. *Xenobiotica* 1, 143.
Goddard, P. & Hill, M.J. (1972). Degradation of steroids by intestinal bacteria. The aromisation of ring A. *Biochimica biophysica Acta*, 280, 336.
Gompertz, D. & Brooks, A.P., Gaya, H. & Spiers, A.S.D. (1973). Volatile fatty acids in the faeces of patients in 'germ-free' isolation. *Gut*, 17, 183.
Gorbach, S.L., Nahas, L., Winstein, L., Levitan, R. & Patterson, J.F. (1967b). The microflora of ileostomy effluent; a unique microbial ecology. *Gastroenterology*, 53, 874.

Gorbach, S.L. (1971). Intestinal microflora. *Gastroenterology*, 60, 1100.
Gorbach, S.L., Banwell, J.G., Mitra, R., Chatterjee, B.D., Jacobs, B. & Mazumaer, D.N.G. (1969). Bacterial contamination of the upper small bowel in tropical sprue. *Lancet*, 1, 74.
Gorbach, S.L., Nahas, L., Lehner, P.I. *et al.*, (1967*a*). Studies of intestinal microflora. I. Effects of diet age and periodic sampling on numbers of faecal micro-organisms in man. *Gastroenterology*, 53, 845.
Gorbach, S.L., Plaut, A.G., Nahas, L . & Weinstein, L. (1967*a*). Studies of intestinal microflora. II. Microorganisms of the small intestine and their relations to oral and faecal flora. *Gastroenterology*, 53, 856.
Gorbach, S.L. & Tabaqchali, S. (1969). Bacteria, Bile and the small intestine. *Gut*, 10, 963.
Gracey, M., Burke, V. & Anderson, C.M. (1969). Association of mono-saccharide malabsorption with abnormal small intestinal flora. *Lancet*, 2, 384.
Gracey, M., Burke, B. & Oshin, A. (1971). Reversible inhibition of intestinal active sugar transport by decon--jugated bile salts *in vitro*. *Biochimica Biophysica Acta*, 225, 308.
Greenberger, N.J., Saegh, S. & Ruppert, R.D. (1968). Urine indican excretion in malabsorptive disorders. *Gastroenterology*, 55, 204.
Hamilton, J.D., Dyer, N.H., Dawson, A.M., O'Grady, F.W., Vince, A., Fenton, J.C.B. & Mollin, D.L. (1970). Assessment and significance of bacterial overgrowth in the small intestine. *Quarterly Journal of Medicine*, 39, 265.
Harries, J.T. & Sladen, G.E. (1972). The effects of different bile salts on the absorption of fluid, electrolytes, and monosaccarides in the small intestine of the rat *in vivo*. *Gut*, 13, 596.
Hawksworth, G. & Hill, M.J. (1971). The formation of nitrosamines by human intestinal bacteria. *Biochemical Journal*, 122, 28 P.
Hepner, G.W., Booth, C.C., Cowan, J. & Mollin, D.L. (1968). Absorption of crystalline folic acid in man. *Lancet*, 2, 302.
Hersh, T., Floch, M.H., Binder, H.H., Conn, H.O., Prizont, R. & Spiro, H.M. (1970). Disturbances of the jejunal and colonic bacterial flora in immunoglobulin deficiencies. *American Journal of Clinical Nutrition*, 23, 1595.
Hill, M.J., Crowther, J.S., Drasar, B.S., Hawksworth, G., Aries, V. & Williams, R.E.O. (1971). Bacteria and aetiology of cancer of large bowel. *Lancet*, 1, 95.
Hill, M.J. & Drasar, B.S. (1968). Degradation of bile salts by human intestinal bacteria. *Gut*, 9, 22.
Hoffbrand, A.V., Tabaqchali, S., Booth, C.C. & Mollin, D.L. (1971). Small intestinal bacterial flora and folate status in gastro-intestinal disease. *Gut*, 12, 27.

Hoffbrand, A.V., Tabaqchali, S. & Mollin, D.L. (1966). High serum folate levels in intestinal blind loop syndrome. *Lancet*, 1, 1339.

Jones, E.A., Craigie, A., Tavill, A.S., Franglen, G. & Rosenoer, V.M. (1968). Protein metabolism in intestinal stagnant loop syndrome. *Gut*, 9, 446.

Kahn, I.J., Jeffries, G.H. & Sleisenger, M.H. (1966). Malabsorption in intestinal scleroderma. Correction by antibiotics. *New England Journal of Medicine*, 274, 1339.

King, R.J.B. (1971). Are oestrogens carcinogens? In: Some Implications of steroid Hormones in Cancer, ed. by D.C. Williams & M.H. Briggs, London: Hienemann, p.62.

Kim, Y.S., Spritz, N., Blum, M., Terz, J. & Sherlock, P. (1966). The role of the altered bile acid metabolism in the steatorrhoea of experimental blind loop. *Journal of Clinical Investigation*, 45, 959.

Knott, F.A. (1927). Addison's (pernicious) anaemia and subacute combined degeneration of the cord: the role of achloryhdria and intestinal infection. *Guy's Hospital Reports*, 77, 1.

Krickler, D.M., & Schrire, V. (1958). 'Kwashiorkor' in an adult due to an intestinal blind loop. *Lancet*, 1, 510.

Lewis, B., Panveliwalla, D., Tabaqchali, S. & Wootton, I.D.P. (1969). Serum bile acids in the stagnant loop syndrome. *Lancet*, 1, 219.

Maier, B.R. & Hentges, D.J. (1972). Experimental shigella infections in laboratory animals. I. Anatagonism by human normal flora components in gnotobiotic mice. *Infection and Immunity*, 6, 168.

McBee, R.H. (1970). Metabolic contributions of the caecal flora. *American Journal of Clinical Nutrition*, 23, 1514.

McLelland, D.B.L., Samson, R.R., Parkin, D.M. & Shearman, D.J.C. (1972). Bacterial agglutination studies with secretory IgA prepared from human gastrointestinal secretions and colostrum. *Gut*, 13, 450.

Mekhjan, H.S., Phillips, S.F. & Hofmann, A.F. (1971). Colonic secretion of water and electrolytes induced by bile acids: perfusion studies in man. *Journal of Clinical Investigation*, 50, 1569.

Meynell, G.G. (1963). Antibacterial mechanisms of the mouse gut. II. The role of Eh and volatile fatty acids in the normal gut. *British Journal of Experimental Pathology*, 44, 209.

Midvedt, T. & Norman, A. (1967). Bile acid transformations by microbial strains belonging to genera found in intestinal contents. *Acta Pathologica et Microbiologica Scandinavica*, 71, 629.

Miller, B., Mitchison, R., Tabaqchali, S. & Neale, G. (1971). The effects of excess bacterial proliferation on protein metabolism in rats with self-filling jejunal sacs. *European Journal of Clinical Investigation*, 2, 23.

Mollin, D.L., Booth, C.C. & Baker, S.J. (1957). The absorption of vitamin B_{12} in control subjects, in Addisonian pernicious anaemia and in the malabsorption syndrome. *British Journal of Haematology*, 3, 412.

Moore, W.E.C., Cato, E.P. & Holdeman, L.V. (1969). Anaerobic bacteria of the gastrointestinal flora and their occurrence in clinical infections. *Journal of Infectious Diseases*, 119, 641.

Neale, G. (1968). Protein deficiency in temperate zones. *Proceedings of the Royal Society of Medicine*, 60, 1069.

Neale, G., Antcliff, A.C., Welbourn, R.G., Mollin, D.L. & Booth, C.C. (1967). Protein malnutrition after partial gastrectomy. *Quarterly Journal of Medicine*, 36, 469.

Neale, G., Gompertz, D., Schjönsby, H., Tabaqchali, S. & Booth, C.C. (1967). The metabolic and nutritional consequences of bacterial overgrowth in the small intestine. *American Journal of Clinical Nutrition*, 25, 1409.

Nelson, D.P. & Mata, L.J. (1970). Bacterial flora associated with human gastrointestinal mucosa. *Gastroenterology*, 58, 56.

O'Gorman, L.P., Borud, O., Khan, I.A. & Gjessing, L.R. (1970). The metabolism of L-3,4-dihydroxyphenylalanine in man. *Clinica Chimica Acta*, 29, 111.

Panveliwella, D., Tabaqchali, S., Wootton, I.D.P. & Lewis, B. (1969). Bile acid metabolism in the stagnant loop syndrome. In: *Proc. 8th International Congress on Nutrition, Prague*. Amsterdam, Excerpta Medica (Int. Congress Series No. 213) p.448.

Parkin, D.M., McClennen, D.B., Percy Robb, I.W. *et al.*, (1970). Intestinal immunoglobulin level and bacterial flora in hypogammaglobulinaemic adult in relation to intestinal absorptive function. *Gut*, 11, 1064.

Percy-Robb, I.W. & Collee, J.G. (1972). Bile acids: a pH dependent antibacterial system in the gut. *British Medical Journal*, 3, 813.

Plaut, A.G., Gorbach, S.L., Nahas, L . *et al.*, (1967). Studies in intestinal microflora. III. The microbial flora of human small bowel intestinal mucosa and fluids. *Gastroenterology*, 53, 868.

Pope, J.L., Parkinson, T.M. & Olson, J.A. (1966). Action of bile salts on the metabolism and transport of water-soluble nutrients by perfused rat jejunum *in vitro*. *Biochimica Biophysica Acta*, 130, 218.

Richards, P., Brown, C.L., Houghton, B.J. & Thompson, E. (1971). Synthesis of phenylalanine and valine by healthy and uraemic men. *Lancet*, 2, 128.
Salen, G., Goldstein, F. & Wirts, C.W. (1966). Malabsorption in intestinal scleroderma. Relation to bacterial flora and treatment with antibiotics. *Annals of Internal Medicine*, 64, 834.
Sandler, M.E., Karoum, F., Ruthven, C.R.J. & Calne, D.B. (1969). m-Hydroxyphenylacetic acid formation from L-dopa in man: Suppression by neomycin. *Science*, 166, 1417.
Savage, D.C. (1970). Associations of indigenous micro-organisms with gastrointestinal mucosal epithelia. *American Journal of Clinical Nutrition*, 23, 1495.
Scheline, R.R. (1968). Drug metabolism by intestinal micro-organisms. *Journal of Pharmaceutic Science*, 57, 2021.
Scheline, R.R. (1973). Metabolism of foreign compounds by gastrointestinal micro-organisms. *Pharmacological Review*, 25, 451.
Schjönsby, H., Drasar, B.S., Tabaqchali, S. & Booth, C.C. (1973). Uptake of vitamin B_{12} by intestinal bacteria in the stagnant loop syndrome. *Scandinavian Journal of Gastroenterology*, 8, 41.
Schjönsby, H. & Tabaqchali, S. (1971*a*). Uptake of rat gastric-juice-bound vitamin B_{12} by intestinal brush borders isolated from blind loop rats. *Scandinavian Journal of Gastroenterology*, 6, 515.
Schjönsby, H. & Tabaqchali, S. (1971*b*). Effect of small intestinal bacteria on intrinsic factor complex. *Scandinavian Journal of Gastrolenterology*, 6, 707.
Scott, A.J. & Kahn, G.A. (1968). Partial biliary obstruction with cholangitis producing a blind loop syndrome. *Gut*, 9, 187.
Smith, H.W. (1966). The antimictobial activity of the stomach contents of suckling rabbits. *Journal of Pathology and Bacteriology*, 91, 1.
Tabaqchali, S. (1970*a*). Case study of a patient with massive intestinal resection. In: *Proc. 7th International Congress Clinical Chemistry, Geneva/Evian 1969*. Vol.4. Basel, Kargar, pp.119-123.
Tabaqchali, S. (1970*b*). The pathophysiological role of small intestinal bacterial flora. *Scandinavian Journal of Gastroenterology, Suppl.* 6. 139.
Tabaqchali, S. & Booth, C.C. (1966). Alteration of bile salt metabolism in the stagnant loop syndrome. *Gut*, 7, 712.
Tabaqchali, S., Hatziaonnou, J. & Booth, C.C. (1968). Bile salt deconjugation and steatorrhoea in patients with the stagnant loop syndrome. *Lancet*, 2, 12.

Tabaqchali, S., Okubadejo, O.A., Neale, G. & Booth, C.C. (1966). Influence of abnormal bacterial flora on small intestinal function. *Proceedings of the Royal Society of Medicine*, 59, 1244.
Tabaqchali, S. & Pallis, C. (1970). Reversible nicotinamide deficiency encephalopathy in a patient with jejunal diverticulosis. *Gut*, 11, 1024.
Van der Reis, V. (1925). Darmbakterien der Erwachsenen und ihre klinische Bedentung. *Ergebn Inn Med. Kinderheilk*. 27, 77.
Vince, A., Dyer, N.H., O'Grady, F.W. & Dawson, A.M. (1972). Bacteriological studies in Crohn's disease. *Journal of Medical Microbiology*, 5, 219.
Vince, A., O'Grady,F.W. & Dawson, A.M. (1973). The development of ileostomy flora. *Journal of Infectious Diseases*. 128, 638.
Wakisaka, G. (1968). Problems in post-operative patients of digestive disease. XI. Blind loop syndrome with megaloblastic anaemia. *Gastroenterology (Japan)*, 3, 109.
Williams, R.T. (1972). Toxicologic implications of biotransformation by intestinal microflora. *Toxicology and Applied Pharmacology*, 23, 769.
Williams, R.C. & Gibbons, R.J. (1972). Inhibition of bacterial adherence by secretory immunoglobulin A: A mechanism of antigen disposal. *Science*, 177, 697.
Yang, M.G., Manoharan, K. & Mickelsen, O. (1970). Nutritional contribution of volatile fatty acids from the caecum of rats. *Journal of Nutrition*, 100, 545.

DISCUSSION

DR SHANSON (UNIVERSITY COLLEGE HOSPITAL): Which antibiotic did you use to decrease the bad effects on absorption of B_{12} in radioactive studies?

DR TABAQCHALI: In fact we used an unusual antibiotic in this situation. As part of an antibiotic trial we gave neomycin in a low dosage of 2 g/day and all it did was reduce the anaerobic flora. The aerobic flora was exactly the same, which was unexpected. This has also been demonstrated with other antibiotics, particularly corrimoxazole, which do not appear to behave *in vivo* as they do *in vitro*. This may be due to the change in the E_h or in the environment of the anaerobic bacteria within the lumen of the small intestine.

DR SELWYN: You mentioned briefly the carcinogens produced by anaerobes but the reverse phenomenen is very fascinating also. This is the apparently protective effect of anaerobes in the bowel against recurrence of cancer. Judging by clinical results there is an increased recurrent rate after resections of large bowel cancers when broad spectrum antibiotics have been given as opposed to narrow spectrum neomycin and so on. I wonder whether any of your work would allow you to speculate on the possible substances produced by the anaerobes that would be particularly toxic to neoplastic cells.

DR DRASAR: There are some animal studies on the treatment of cancers using clostridia with the spores locating in the cancer and then the clostridial toxins, because of the anaerobic conditions in the cancer, destroying the carcinoma cells. I wouldn't like to say that that was of any importance in human beings, nor would I like to advocate it as a treatment.

DR OKUBADEJO (PORTSMOUTH): I would like to ask Dr Tabaqchali about the uptake of Vitamin B_{12} combined with intrinsic factor by bacteria and whether if you incubate long enough, the uptake that you couldn't demonstrate would in fact happen. I know of a piece of work in which this was said to be so and I wonder if you could either confirm or say whether it is not so.

DR TABAQCHALI: It has been shown that if you incubate bacteria for over 48-72 hours there is a certain amount of uptake of intrinsic factor bound Vitamin B_{12} by a variety of micro-organisms. In fact the ones tested were *Pseudomonas* spp., *Escherischia coli* and *Proteus* spp. but the uptake is not very much; it may only be 1 per cent. It is a very slight uptake which is very different from what we have shown in our studies with certain strains of *Bacteroides* which appeared to take up about 10-15 per cent of bound B_{12}.

GENERAL DISCUSSION ON LABORATORY ASPECTS OF ANAEROBIC INFECTIONS

DR BOSWELL (KING'S COLLEGE HOSPITAL): Dr Phillips mentioned, very briefly, cephalosporins in his antibiotic sensitivity testing and said that some of them were probably broken down by enzymes. Is it just lack of time that has stopped him producing the cephalosporins on his chart?
DR PHILLIPS: Yes it is. We have some other results on, I think, three cephalosporins but we're just working on it at the moment and so I can't go further than I've already gone on that one.
DR NEWSOM (PAPWORTH HOSPITAL): Firstly, could you tell me what happens to the *Bacteroides* in the normal gut when you give an aminoglycoside to a patient? Secondly, I noticed that the last speaker talked about rats and indole and *Bacteroides*. I did not grow any *Bacteroides* from the few rat samples that I have tested, and I wondered if they are present in the normal rat gut flora.
DR TABAQCHALI: I would be prepared to comment on the second question. We studied the bacteriology at different levels of the small intestine of the normal rat and also in rats with experimental blind loops. In fact *Bacteroides* are present in both situations and in the blind loop they go up to 10^9/ml.
DR WATT: Could I perhaps comment on the first point as a clinical microbiologist. It has always seemed to me, with all respect to Dr. Tabaqchali, that giving neomycin to people for a bowel preparation must encourage the suppression of other flora. As we use neomycin blood agar as a selective agent it may well be that we produce a similar selection in the patient. I think this is borne out by the experience that postoperative *Bacteroides* infections, certainly in the laboratory in which I work, are often associated with patients who have had preoperative bowel preparation with neomycin and we can often isolate *Bacteroides* from patients who are receiving large doses of, for instance, gentamicin.
DR COLLEE: Some people feel very strongly about preoperative bowel preparation but fortunately we may be speaking predominantly to physicians rather than surgeons today, otherwise we would certainly have some more agitated conversation.
DR PERCIVAL: We have counted the *Bacteroides* in people with leukaemia on oral gentamicin (100 mg tds) and although the aerobic enterobacteria were reduced to virtually nil, there was no drop at all in the number of *Bacteroides*.
DR COLLEE: These were faecal counts were they Dr Percival?
DR PERCIVAL: Yes, sir.

DR SLEIGH (GLASGOW): Could the experts advocate any transport medium to assist the survival of anaerobes in clinical microbiology?

DR DRASAR: The important thing about any transport medium or transport system for anaerobes is that it should be anaerobic and also prevent the drying out of the specimen. The use of gas tubes, gassed at the bedside, has been advocated, as has the use of reducing systems and various small anaerobic jar-type systems. I think probably the most readily available medium that could be used as a transport medium would be Robertson's meat broth.

DR COLLEE: I think that, as your Chairman, I should call upon Dr Shanson to talk to that because he has clear ideas on thioglycollate and I know that Dr Finegold has clear ideas on transport of specimens. Then perhaps I might add a point. The question relates to the carriage of specimens from the patient to the laboratory and you might like to talk specifically about blood culture which is a good one to start with.

DR SHANSON: Could I first make a comment on the cooked meat. I would have thought that one runs into the risk there of overgrowth of aerobic organisms for example and that this wouldn't be suitable for clinical purposes. In practice the use of Stuart's transport medium probably meets the problem, but I know of no scientific work on this. As regards blood cultures, the use of thioglycollate medium, in my experimental experience, gives better results than cooked meat methods but the use of a particular thioglycollate is important. If one uses, for example, a thioglycollate method with the United States Pharmacopoeia formula, one may get very rapid growth of the anaerobes but unfortunately one may also get rapid killing of the organisms in this medium. If the laboratory subcultures its blood cultures routinely once a week, say, it is possible that the organisms may be dead whereas if one uses a suitable type of Brewer's thioglycollate this risk seems to be less. A paper is in press on this which will describe the particular Brewer's that appeared to give the best results and will be published soon in the Journal of Clinical Pathology. One last point about the blood culture methods, is that liquoid does seem to inhibit some strains of anaerobic cocci and therefore, the method that I would prefer is the use of a suitable thioglycollate Brewer's medium without liquoid. I would suggest the use of 80 ml in a three ounce bottle; if 5 ml of blood are added one gets round the anti-coagulant problem and also the antibacterial effect of fresh blood.

DR COLLEE: Thank you very much for these helpful remarks Dr Shanson. Dr Finegold, may we tempt you to talk a little about the transport of specimens?

DR FINEGOLD: I like the approach that Dr Drasar has recommended, that is, to use a gassed-out tube. We do like to have a little pre-reduced medium in there, if only as a vehicle for an oxidation-reduction indicator, because we find that these tubes are tricky and no matter how well you think you have the gassing procedure done either they don't get adequately anaerobic in the first place or they may, despite the use of butyl rubber stoppers and so on, become aerated later on. I think it's important to incorporate resazurin in the tube in one form or another to be sure it's purely anaerobic when you put your specimen in. Dr Drasar also mentioned a miniature anaerobic jar which is actually a 35 mm film container utilizing the steel wool copper sulphate technique for achieving anaerobiosis. This works very nicely for tissue specimens and I think is a preferred method if one must use swabs, although it's a little bit cumbersome. The other thing that can be done with swabs is to use a semi-solid transport medium and Stuart's transport medium is satisfactory in the sense that anaerobes will survive in this for extended periods of time. The problem, however, is that facultative forms will utilize the glycerol phosphate in Stuart's and will grow, whereas the anaerobes will remain static, and so you'll distort your numerical relationships or, perhaps, even kill the anaerobes off in time. We think that similar media, such as Cary and Blair medium which does not contain glycerophosphate would probably be the preferable approach if one must use a swab although we haven't specifically studied that. Finally, with regard to the question of Liquoid, there is another similar compound, sodium amylosulphate (not yet available commercially to my knowledge, but apparently soon to be available), which gets around the drawback which was mentioned with regard to sodium polyanetholsulphonate - that is, of inhibition of certain anaerobes, specifically certain of the anaerobic streptococci. Sodium amylosulphate apparently has all of the beneficial characteristics of Liquoid with none of the drawbacks.

DR COLLEE: Thank you very much Dr Finegold. In some studies we have done which will shortly be published, with admittedly contrived models which will need to be subjected to clinical experience, we would put three factors in the following order of importance. First, let your specimen be wet, second let it be prompt and third if it should be reducing then fair enough.

DR NAKHLA (ST ALBANS): Depending on the clinical situation, if we are really hunting for *Bacteroides*, we can just plate them at the bedside, as Dr Percival said and you can increase the isolation rate from 20 to 80 per cent.

DR WILLIS: The fact that we want a specimen quickly for anaerobic work implies realisation by clinical colleagues that this may be a specimen which contains anaerobes. I suspect we may come back to this problem later. One of our difficulties is, although we think we're looking at all the specimens we think may contain anaerobes, it's not uncommon to get specimens such as swabs where the request is 'orgs and sens.' and there is no indication really where the swab is from; it is quite impracticable of course to ring up about each swab we get. I think this is important. There must be a clinical awareness and perhaps a suggestion that anaerobes are being looked for. The problem of a transport medium perhaps may lull the porters of the hospital into a state of shock so that the specimens will take 24 hours to reach us and it may even lull the technicians into a state of inactivity because it is in the transport medium so that not only is there a loss of time in the specimen reaching the laboratory but then it's not handled promptly. I think personal handling of these specimens by, preferably, a clinical microbiologist in consultation with the clinical staff concerned is really essential for considerable success.

DR COLLEE: Thank you for that timely warning that transport media are often delayed.

DR GOODWIN (NORTHWICK PARK): I think there probably is a difference in the epidemiology of sulphonamide-resistant *Bacteroides*. I don't know whether the pre-operative treatment in Portsmouth and other places is different but if sulphonamide is not used preoperatively then it is quite likely that the proportion of *Bacteroides* which are sulphonamide-sensitive will be much higher.

I wonder whether the incidence of *Bacteroides* in Dr Tabaqchali's upper bowel specimens is 100 per cent in the patients clinicians send her? I find this a very unrewarding experience - they send them along in large numbers expecting us to find *Bacteroides* and we don't find them as often as they would like.

DR TABAQCHALI: No I think youre absolutely right because, as I said, the type of bacteria and their concentrations depend on the type and extent of the causative intestinal lesion. There are numerous conditions where you don't isolate anaerobes, as in uncomplicated gastrectomy cases, in which you may just find low counts of enterobacteria throughout, but no anaerobic bacteria. I think it depends on the type of lesion and the degree of stasis.

DR BOSWELL (KING'S): Having gone to all this trouble to isolate these organisms, could you recommend to us the best way to keep them alive in the laboratory while we work with them?

DR WATT: We find that keeping them in cooked meat broth is satisfactory with the proviso that they do need subculture at fairly regular intervals and obviously this means checking for purity. Dr Drasar might elaborate on possible ways of deep freezing them.

DR DRASAR: Most of these organisms will deep freeze in culture media if one adds sterile glycerol to a concentration of about 10 per cent. Most satisfactorily one will add the glycerol to the media and then freeze the culture immediately in liquid nitrogen. Addition of glycerol to Robertson's meat broth and putting it in a deep freeze keeps organisms satisfactorily for about 6 months. One loses a few but if they've grown well, you can usually recover them.

DR COLLEE: Of course this is proper cooked meat broth; there are cooked meat broths on the market that are not cooked meat in the normally accepted sense and these are excluded, I think, from your remarks Dr Drasar.

DR SLEIGH (GLASGOW): May I ask, Sir, is that just at -20° or at -70°?

DR DRASAR: It's better at -70° but does work at -20°.

DR OKUBADEJO (PORTSMOUTH): Can I ask about the bacteroidicine effect? I was wondering whether the bacteroidicine produced would inhibit the growth of other *Bacteroides* or whether it is just *Bacteroides fragilis*.

DR COLLEE: With 50 strains that we looked at representing different species of *Bacteroides*, we found that there was a cross-over effect, a non-specific effect if that is the right word, between producers and indicators. For instance *Bacteroides fragilis* produced effects that were noticeable with *necrophorus*-like strains and *melaninogenicus*-like strains. In the very limited number of other tests that we have done, I don't think we have an effect active against a *Fusobacterium*-like strain and the effect was very specific in terms of its not being effective against a wide range of other organisms that we tested. There was an unusual effect noticeable with *Streptococcus pneumoniae* which may be a nutritional effect, but I'm not sure.

DR GHONEIM (EXETER): Are there any studies on the antigenic structure of *Bacterides fragilis* to confirm its uniformity?

DR OKUBADEJO: With phenol-extracted lipopolysaccharide from various strains of *Bacteroides fragilis* nearly all of them give us the same antigenic cross-over and we can't really distinguish between them at the moment except for one strain out of about 50.

DR COLLEE: Thank you for that cautious reply. The literature in this field is fairly confused at the moment but there does seem to be a point worthy of follow up in relation to serological analysis of this difficult group.

CLINICAL ASPECTS OF ANAEROBIC INFECTIONS

6. Anaerobes in the Mouth

J. HARDIE

The significance of anaerobic bacteria as part of the normal commensal flora, and of their role in a variety of infections, has become more widely appreciated in recent years (Drasar, 1967; Moore, Cato & Holdeman, 1969; Zabransky, 1970; Finegold & Rosenblatt, 1973; Levison, 1973).

Improved isolation and identification techniques which have been described for anaerobes (Killgore *et al.*, 1973) are only just beginning to be applied to the study of the oral microflora. Most of the currently available literature on oral anaerobes is based upon studies in which traditional anaerobic techniques have been employed. It is probable that some of our current ideas on the role of anaerobes in the mouth, and in oral infections in particular, will need to be revised in the light of information gathered by means of modern techniques.

The fact that the human mouth harbours large numbers of anaerobes has been known for many years (see Rosebury, 1962; Burnett & Scherp, 1968; Nolte, 1973; for general reviews). Information on methods of isolation and characterization of oral anaerobic bacteria has been summarized previously by Bowden & Hardie (1971). The excellent handbooks by Holdeman & Moore, (1972) and Sutter *et al.*, (1972) provide detailed descriptions of many of the methods available for laboratory investigations on anaerobes and have been found to be extremely useful in our laboratory. However, it is not unusual to isolate anaerobic bacteria from the mouth that do not seem to conform to published descriptions and a number of taxonomic problems remain to be resolved.

In this chapter a brief account will be given of the types of anaerobic bacteria that are normally present in the mouth, their distribution in different sites, and the relationship of anaerobes to oral infections.

ANAEROBIOSIS IN THE MOUTH

The importance of anaerobiosis as an ecological determinant in the mouth has been reviewed recently (Hardie & Bowden, 1974) and only a few of the salient points will be reiterated here. Studies on the microbial flora of the gingival crevice area by Socransky *et al.*, (1963) indicated that only 20-25 per cent of the total microscopic count could be cultivated by means of conventional anaerobic techniques. Using the Hungate (1966) roll-tube technique, Gordon, Stutman & Loesche, (1971) were able to increase the recovery of viable organisms from gingival samples to about 70 per cent of the microscopic count. Similarly, the use of an anaerobic chamber technique also allows the isolation of a much higher proportion of anaerobic bacteria from oral samples (Aranki *et al.*, 1969). It appears from these investigations that the critical stage in the laboratory analysis of samples from the mouth occurs during collection, dispersion, dilution and plating procedures. Later in this review a few studies will be referred to in which more exacting anaerobic conditions have been employed for examination of clinical specimens, and in each case a much higher recovery of anaerobes has been recorded than that obtained conventionally.

The oxygen sensitivity of some anaerobic bacteria has been examined by Loesche (1969), and some of his observations on oral organisms are summarized in Table 6.1. The most

Table 6.1. Oxygen sensitivity of certain anaerobic bacteria (adapted from Loesche, 1969).

Strict anaerobes (maximum growth at $pO_2 < 0.5\%$)	Moderate anaerobes (maximum growth at $pO_2 < 3\%$)	Microaerophils (maximum growth at $pO_2 > 3\%$ and $< 20\%$)
Treponema macrodentium (1)*	*Bacteroides fragilis* (7)	*Vibrio sputorum*
Treponema denticola (2)	*B. melaninogenicus* (5)	var. *sputorum* (2)
Treponema oralis (1)	*B. oralis* (4)	var. *bubulum* (2)
Selenomonas ruminatium (1)	*Fusobacterium nucleatum* (3)	*Vibrio fetus* (1)
	Peptostreptococcus elsdendii (1)	

* Numbers of strains tested in parentheses

oxygen sensitive group of organisms included several *Treponema* strains, whilst *Bacteroides*, *Fusobacterium* and *Peptostreptococcus* species fell into a group described as moderate anaerobes. The latter group could grow in the presence of 2-8 per cent oxygen and were able to survive up to 90 minutes exposure to atmospheric oxygen.

Eskow & Loesche (1971) measured the oxygen tension in different parts of the mouth and their findings are reproduced in Table 6.2. The levels of 0.4 per cent and 0.3 per cent recorded for the maxillary and mandibular buccal folds respectively are probably low enough to allow growth of many anaerobic species. The oxidation-reduction

Table 6.2. Mean oxygen concentration in the various areas of the oral cavity* (from Eskow & Loesche, 1971).

Area	% Oxygen
Maxillary buccal fold	0.4 ± 0.1
Mandibular buccal fold	0.3 ± 0.1
Anterior tongue	16.4 ± 0.2
Posterior tongue	12.4 ± 0.7

* Values obtained from 20 subjects.

± Standard error of the means.

potentials of the gingival crevice area have been measured by Kenney & Ash (1969) and their results are shown in Table 6.3. A significant difference was observed between the eH of normal gingival sulci and that of deeper periodontal pockets. The same authors also monitored the changes in eH

Table 6.3. Oxidation-Reduction potentials in the gingival area (after Kenney & Ash, 1969).

Subjects	Areas examined	Mean O-R Potential (mV)	Range
Patients with periodontal disease	Periodontal pockets	- 47.6	-157-+14
	Gingival sulci	+ 72.6	+10-+113
Control subjects	Gingival sulci	+ 74.4	+33-+103

that occur during the development of dental plaque and the readings obtained from two subjects are illustrated in Table 6.4. These observations clearly demonstrate that the change from a relatively simple flora to the highly complex situation that exists in older plaque is accompanied by a significant change in the oxidation-potention from over +244 millivolts to about -141 mV after about two weeks development. Such a reduced environment in the gingival crevice is highly suitable for the growth of the more fastidious anaerobes.

Table 6.4. Oxidation-Reduction potentials of developing plaque (adapted from Kenney & Ash, 1969).

Days	Oxidation-Reduction Potential (mv) Subject 1	Subject 2	Bacteria seen in stained smears
0	+ 244	+ 294	Gram-positive cocci and rods
1	+ 63	...	Gram-positive cocci
2	...	- 30	Gram-positive cocci and rods Gram-negative cocci and rods
3	- 31	- 67	As above
4	- 112	...	As above
7	...	- 141	Complex flora

THE TYPES OF ANAEROBES FOUND IN THE MOUTH

The principal anaerobic or micro-aerophilic genera known to occur regularly in the oral cavity are listed in Table 6.5. Much of the literature on the properties of the organisms listed in this table has been reviewed in recent publications (Bowden & Hardie, 1971; Bowden & Hardie, 1973; Hardie & Bowden, 1974).

Table 6.5. Non-sporing anaerobes found in the mouth.

Cocci		Rods and filaments	
Gram-positive	Gram-negative	Gram-positive	Gram-negative
Peptococcus	*Veillonella*	*Actinomyces*	*Bacteroides*
Peptostreptococcus		*Eubacterium*	*Fusobacterium*
		Leptotrichia	*Campylobacter (Vibrio)*
		Arachnia	
		Propionibacterium	
		Bifidobacterium	(*Spirochaetes*)

COCCI

The anaerobic Gram-negative cocci belong to the genus *Veillonella*, and present few problems for isolation or identification. They are often present in relatively large numbers and can be isolated from most parts of the mouth. *Veillonella* species almost invariably constitute a significant proportion of the flora of dental plaque.

Facultative streptococci abound in the mouth and have been extensively studied by many workers. On the other hand, strictly anaerobic Gram-positive cocci of the genera *Peptococcus* and *Peptostreptococcus* have received much less attention although they have been reported in several studies. In our own investigations on dental plaque from children's mouths we rarely isolate these organisms, although other fastidous anaerobes are regularly obtained. However, in a recent study employing continuous anaerobiosis during collection and handling of oral samples, Berg & Nord (1972) have shown them to be present in developing dental plaque. It appears that these anaerobic cocci are particularly sensitive to oxygen exposure during collection and dispersion of samples, and more care is required for their regular isolation.

GRAM-POSITIVE RODS AND FILAMENTS

This group of bacteria comprises a wide variety of genera and species which together often constitute a substantial proportion of the oral microflora. Up to 50 per cent or more of the total cultivable flora of dental plaque may consist of Gram-positive rods, often with *Actinomyces* species predominating (Hardie & Bowden, 1974; Bowden, Hardie & Slack, 1974). Most of the published studies on dental plaque have shown that Gram-positive rods are as numerous as streptococci, although the latter have been more extensively investigated.

Identification of anaerobic and facultative Gram-positive rods and filaments often poses problems in the laboratory, although progress has been made in the taxonomy of some of these genera (Pine, 1970; Bowden & Hardie, 1973; Cross & Goodfellow, 1973; Holmberg & Hallander, 1973; Holdeman & Moore, 1972). It should be possible, with the information currently available, to go beyond the stage of calling all such isolates 'anaerobic diphtheroids'. Identification of metabolic products by gas-liquid chromatography is one extremely useful technique for characterization of these organisms (Holdeman & Moore, 1972), and a simplified method has recently been described by Carlsson (1973).

All of the organisms listed in Table 6.5 can be isolated with good conventional anaerobic methods, with pre-reduced culture media. However, it is probable that the isolation rate for some species would be considerably increased if all samples were handled in an anaerobic chamber. Although *Propionibacterium*, *Arachnia*, *Bifidobacterium* and *Eubacterium* species have been found in dental plaque and deep dentinal caries, there are insufficient data available at present to estimate their frequency of occurrence or numbers in the mouth.

Clostridia do not seem to be regular inhabitants of the mouth in most of the communities that have been studied, although they have been recorded by some workers (Van Reenan & Coogan, 1970). Loesche, Hockett & Syed, (1972) found that clostridia constituted about 8 per cent of the total cultivable flora of dental plaque collected from institutionalized subjects. This observation was quite unexpected, since numerous other investigations on plaque have failed to disclose the presence of any *Clostridium* species. It is possible that faecal contamination of the mouth, which may occur in mentally retarded subjects, could lead to the establishment of a different bacterial flora.

GRAM-NEGATIVE RODS AND FILAMENTS

Most published reports in which oral *Bacteroides* have been characterized indicate that the prevalent species are *B. melaninogenicus*, *B. oralis* and *B. corrodens*. Perhaps surprisingly, there does not appear to be much information on the occurrence of *B. fragilis* in the mouth, and it is not possible to state with certainty whether this species is a regular oral inhabitant. Studies in progress (Williams, Bowden, Hardie & Shah, *unpublished data*) indicate that black-pigmented *Bacteroides* species isolated from the mouth differ from those normally obtained from the gut, in several respects, including cell wall composition and glucose utilization. The possible role of *Bacteroides* in disease processes in the mouth is referred to in a later section of this chapter.

Fusobacteria from the mouth are usually identified as *F. nucleatum* or *F. polymorphum*. Other fusiform types may also be present, but few investigators have studied them in any detail. The numbers of fusobacteria in saliva and gingival material have been studied by Hadi & Russell (1968, 1969). *Leptotrichia buccalis* often gives a similar appearance to fusobacteria in stained smears, although it is usually Gram-positive in young cultures and presents a different colonial morphology.

Little attention has been paid to anaerobic vibrios in the mouth, such as *Vibrio sputorum* (Loesche, Gibbons & Socransky, 1965) (or *Campylobacter*, Veron & Chatelain, 1973). These organisms have been shown to be microaerophilic rather than strictly anaerobic (Loesche, 1969). The role of these and other motile anaerobic organisms such as *Selenomonas* in oral conditions is not well understood. Loesche & Gibbons (1965) have described a simple identification scheme for the most numerous oral Gram-negative anaerobic rods.

SPIROCHAETES

The oral cavity has been known for many years to harbour several types of spirochaetes, and Socransky *et al.*, (1963) have estimated that gingival debris may contain 1.7 x 10^9 of these organisms per gram wet weight. Hampp and his co-workers have conducted extensive investigations on methods of isolation, identification, pathogenicity and other properties of oral spirochaetes (Hampp, 1947, 1957; Hampp & Mergenhagen, 1961, 1963). Much of the early literature has been reviewed by Rosebury (1962), and some of the laboratory methods available for studying spirochaetes have been reviewed (Bowden & Hardie, 1971).

The electron microscope has been used quite extensively for characterization of oral spirochaetes. Listgarten & Socransky (1965) divided a collection of strains into three groups on the basis of their size, and the number and arrangement of axial fibrils. The group of small spirochaetes was subsequently differentiated into three species, namely *Treponema denticola, T. macrodentium* and *T. oralis*, by means of morphological and biochemical criteria (Socransky *et al.*, 1969).

The difficulty experienced in isolating oral spirochaetes in pure culture has inhibited most workers from conducting cultural studies on these organisms. Hopefully, the current trend for investigating anaerobes in various infections will lead to a re-evaluation of the role of spirochaetes in oral diseases, particularly acute ulcerative gingivitis (AUG) and chronic periodontitis.

THE DISTRIBUTION OF ORGANISMS IN THE MOUTH

Different parts of the mouth appear to favour the growth of some bacteria in preference to others, with the result that the proportion of the total flora made up by particular species varies from one site to another (Socransky & Manganiello, 1971, Davies, 1972). The ecological situation which exists in the oral cavity is complex, and the micro-environment changes from time to time with the eruption and loss of teeth, alterations in diet, development of dental disease and numerous other factors. Site variations in the composition of the oral flora have been discussed at some length recently (Hardie & Bowden, 1974), and the general pattern of distribution of bacteria in saliva, tongue, supra-gingival dental plaque and gingival crevice material is summarized in Table 6.6.

Anaerobic bacteria of various types can be isolated from all parts of the mouth, but are usually present in the highest numbers in dental plaque, especially that which develops in the gingival crevice region. The mouth of a new born infant rapidly becomes colonized by micro-organisms, and anaerobic species can be isolated at an early age, increasing in numbers as the child grows older (McCarthy, Snyder & Parker, 1965). When dental plaque develops on a clean tooth surface, numerous studies have shown that the earliest colonizing organisms are predominantly aerobic and facultative. Within a few days anaerobic species increase in numbers and may become dominant after 10-14 days (Slack & Bowden, 1965; Ritz, 1967). This progressive development of the tooth surface flora is accompanied by a drop in eH, as previously noted (Kenney & Ash, 1969).

The predilection of certain anaerobes for the gingival crevice area, especially *Bacteroides* species and spirochaetes, has been noted by several authors (Gibbons *et al.*, 1963; Dwyer & Socransky, 1968; Carlsson, 1971). Presumably the combination of a suitably low eH, together with the availability of appropriate nutritional requirements (Loesche, 1968) provides the ideal conditions for proliferation of strict anaerobes.

THE RELATIONSHIP OF ANAEROBES TO ORAL INFECTIONS

Most of the bacterial infections that occur in the mouth are thought to be caused by members of the normal commensal flora. The two most widespread oral diseases,

Table 6.6. Mean percentages of cultivable bacteria in the adult oral cavity (adapted from Socransky & Manganiello, 1971).

Micro-organisms	Tongue	Saliva	Gingival Crevice	Dental Plaque
Gram-positive anaerobic cocci	4.2	13.0	7.4	12.6
Gram-positive anaerobic rods	8.2	4.8	20.2	18.4
Gram-negative anaerobic cocci	16.0	15.9	10.7	6.4
Gram-negative anaerobic rods	8.2	4.8	16.1	10.4
Spirochaetes	N.D.	N.D.	1.0	N.D.
Gram-positive facultative cocci	44.8	41.0	28.8	28.2
Gram-positive facultative rods	13.0	11.8	15.3	23.8
Gram-negative facultative cocci	3.4	1.2	0.4	0.4
Gram-negative facultative rods	3.2	2.3	1.2	N.D.

N.D. = not detected.

dental caries and periodontal disease, are directly attributable to the damaging effects of dental plaque. The importance of the commensal flora in opportunist infections in various parts of the body has been reviewed recently by Klainer & Beisel, (1969), Finegold, Marsh & Bartlett, (1971), and Williams, (1973).

In this section, the role of anaerobes in various pathological conditions will be discussed, and some of the more recent literature reviewed. General descriptions of the clinical and other aspects of oral infections may be found in standard textbooks such as those by Burnett & Scherp, (1968); Nolte, (1973) and Killey, Seward & Kay, (1971).

DENTAL CARIES

Most of the experimental evidence concerning the microbial aetiology of dental caries indicates that streptococci, particularly *Streptococcus mutans*, are the agents chiefly responsible for initiating the disease. A few other organisms, such as strains of *Lactobacillus* and *Actinomyces*, have also been shown to be capable of inducing caries in experimental animals and there is some data to suggest that *A. naeslundi* and *A. viscosus* may be involved in root surface caries in man (Jordan & Hammond, 1972; Jordan & Sumney, 1973).

There is little reason to suppose at present that strict anaerobes play a vital part in the initiation of dental caries. However, Mikx *et al.*, (1972) have demonstrated that *Veillonella* species can modify and reduce the cariogenic potential of *Strep. mutans* in an animal model system. The mechanism for this caries reducing effect is not fully understood, but may be linked to the ability of *Veillonella* to utilize lactic acid. Thus, although the anaerobic bacteria in plaque are probably not responsible for initiating damage to sound enamel, their presence may have an important effect in modifying the destruction caused by other organisms under certain conditions.

Once the tooth surface has been breached, the resulting cavity harbours a variety of microbial species. Streptococci, lactobacilli, and actinomyces have received particular attention in this context, but only recently have attempts been made to isolate anaerobes by the more modern techniques now available. Edwardson (*personal communication)* has investigated the bacterial flora of the advancing front of carious lesions that have reached the dentine, using freshly extracted teeth. In this study, contamination by organisms at the tooth surface was avoided by approaching the lesion from the sterile pulpal aspect and working outwards until positive cultures were obtained. The isolates obtained from 46 teeth examined in this way have been characterized and placed into five main groups as shown in Table 6.7.

Table 6.7. The prevalence of different bacterial groups isolated from deep dentinal carious lesions in 46 teeth (data from Edwardsson, 1974 - *personal communication*).

Group	Organisms	Per cent of teeth in which group was isolated	Per cent in pure culture
I*	*Actinomyces*	72	26
	Bifidobacterium		
	Eubacterium		
	Propionibacterium		
	Corynebacterium		
II*	*Lactobacillus*	48	26
III	*Streptococcus*	18	4
IV	*Veillonella*	9	0
	Selenomonas		
	Bacteroides		
V*	Other anaerobic Gram-positive rods and cocci	19	4

* A detailed taxonomic study on the anaerobic Gram-positive rods and filaments isolated in this investigation will be published shortly (Edwardsson, S., & Bowden, G.H., *in preparation*).

The predominant organisms cultured from the advancing front of these lesions were the Gram-positive rods of Group I, together with lactobacilli (Group II). Streptococci were obtained in pure culture from two samples, but altogether were isolated from only 8 (17 per cent) of the specimens. The organisms of Groups IV and V were only occasionally isolated in pure culture, but occurred in mixed culture in several of the teeth examined. From this very carefully conducted study it appears that anaerobic bacteria may play a significant role in the progression of caries from the initial enamel lesion into the deeper dental tissues. Edwardsson's work indicates that Gram-positive organisms predominate in deep dentinal caries, since very few Gram-negative bacteria were isolated in the study just described.

A similar low isolation rate of anaerobic Gram-negative organisms from carious dentine has been reported by Loesche and Syed, (1973). In their study, low numbers of *Veillonella* were isolated but no anaerobic Gram-negative rods or filaments were recorded.

INFECTED PULPS, ROOT CANALS AND DENTAL ABSCESSES

Infections of the pulp and the periapical region usually arise either by direct extension of dental caries into deeper tissues or follow trauma to the tooth. Teeth with dead and infected pulps can be preserved by endodontic treatment, the object of which is to render the root canal free of micro-organisms by mechanical and chemical means prior to the insertion of a root filling. Adequate sterilization of the root canal together with efficient sealing usually allows the resolution of any residual periapical infection and repair of bony defects around the apex of the treated tooth.

Bacteriological control of endodontic therapy is desirable, and this is one of the few occasions when regular use is made of bacteriology reports during dental treatment. Many investigations have been carried out upon the microbial flora of the infected root canal, and the extensive literature has been reviewed by Möller, (1966). Until comparatively recently little attention has been devoted to the involvement of anaerobes in these infections. The organisms most frequently reported have been streptococci and staphylococci,together with various other aerobic and facultative genera.

In his comprehensive methodological studies, Möller, (1966) clearly demonstrated that improved laboratory methods for examining root canal samples resulted in a significantly higher proportion of positive cultures than was previously achieved. This increase was attributable almost entirely to an improved recovery of anaerobes. The bacteria most commonly isolated from apical areas were streptococci, lactobacilli and a variety of anaerobes. The predominant anaerobic types recovered were Gram-positive cocci (*Peptococcus* and *Peptostreptococcus*) and Gram-positive rods (*Eubacterium*, *Lactobacillus* and *Corynebacterium*). Möller concluded his thesis by stating: 'These findings indicate that, in addition to the facultative anaerobic streptococci commonly found, the strictly anaerobic, non-sporulating bacteria may be of importance for the pathological changes of the apical area'.

Recently, two short reports have been published which indicate that more exacting anaerobic methods will reveal even more anaerobes from necrotic pulps and root canals. In a pilot study by Fulghum, Wiggins & Mullaney, (1973), a comparison was made between conventional techniques, using thioglycollate medium, and the VPI system, in which pre-reduced medium 'E' was employed for incubation of root canal samples. Duplicate samples from 24 specimens yielded 18 (75 per cent) positive cultures in the medium 'E', whereas only 10 (42 per cent) grew in the thioglycollate medium. The identity of the bacteria grown was not established, but the results indicated that a higher isolation rate could be obtained by the use of more strict anaerobiosis.

Berg & Nord, (1973) have also described a method for isolating anaerobes from endodontic specimens. In their study the samples were collected from the tooth under a continual stream of oxygen-free gas which was delivered from a mobile unit at the chairside. The gas was channelled through specially adapted rubber dam clamps and tweezers, so that the specimen was not exposed to atmospheric air. The samples obtained in this way were immediately transferred to pre-reduced anaerobic roll-tubes, all manipulations being performed under oxygen-free gas.

The authors examined samples from 20 teeth and compared this technique with a more conventional sampling method in which the specimens were placed in VMG III transport medium (Möller 1966). In 11/20 teeth, the special anaerobic collection method allowed the isolation of anaerobic species which were not recovered from the duplicate samples examined conventially. The organisms found included *Bacteroides*, *Veillonella*, *Bifidobacterium*, *Peptostreptococcus*, *Peptococcus*, and *Propionibacterium* species.

A bacteriological study of dental abscesses by Feldmann & Larje, (1966), in which careful sampling techniques were used, showed that anaerobic organisms were frequently present, although 'viridans-type' streptococci predominated. Pyogenic staphylococci occurred in only two of the 73 cases examined, and no β-haemolytic streptococci were found. These authors reported a higher incidence of anaerobic streptococci, *Veillonella*, *Fusobacterium* and *Leptotrichia* than that described by previous investigators, and they also emphasized the importance of using adequate anaerobic methods for the examination of pus from dental abscesses.

Moore & Russell, (1972) also investigated the bacteriology of dental abscesses and showed that aspiration was preferable to swabbing for obtaining reliable samples which were relatively free from contamination by saliva. No particular emphasis was placed upon the recovery of anaerobes in this study, although a few isolations of fusobacteria and anaerobic streptococci were recorded. However, it is probable that the three day primary isolation period used was insufficient to allow detection of some of the slower growing anaerobes which may have been present.

Infection of dental origin may spread to involve deeper tissues, resulting in the condition known as Ludwig's angina (Fein & Mohnac, 1973). The involvement of anaerobes in this infection, although likely, is not well documented at present.

CHRONIC PERIODONTAL DISEASE

Chronic gingivitis is an inflammatory condition of the gingivae, which if unchecked, commonly advances to involve the periodontal membrane and alveolar bone. This process, known as chronic periodontitis, is characterized by progressive loss of the supporting tissues of the teeth and is the main cause of tooth loss in adults. Virtually the whole of the world's population is affected to some degree by periodontal disease.

All the available evidence indicates that there is a direct relationship between the extent of periodontal disease and the amount of dental plaque in the mouth (Greene, 1963; Russell, 1963; Sheiham, 1970). The fact the older people tend to have more severe periodontal disease than younger subjects is probably related to the length of time the periodontal tissues have been exposed to the irritant effects of plaque (Sheiham, 1970). Short-term studies on experimental gingivitis in man have shown that the onset of gingivitis is related to the accumulation of the typical complex plaque flora which develops within 7-10 days after withdrawal of toothbrushing (Theilade *et al.*, 1966; Loe, Theilade & Jensen, 1965). At this stage of development the plaque contains an abundance of anaerobes, including spirochaetes and various filamentous organisms.

Although it is quite clear that dental plaque is responsible for initiating periodontal disease, it is not possible to state at present which particular micro-organisms are essential to the destructive process. The extensive literature concerning the relationship of plaque bacteria to periodontal disease has been reviewed recently by several authors, including Genco, Evans & Ellison, (1969), Bahn, (1970) and Socransky, (1970), and no attempt will be made to cover all the same ground again here. *Bacteroides melaninogenicus* has attracted some attention in several studies on the microbiology of periodontal disease. This organism, as already mentioned, is more easily isolated from the gingival crevice than from other areas in the mouth, and its numbers apparently increase with age (Bailit, Balwin & Hunt, 1964; Kelstrup, 1966). Studies on the pathogenesis of mixed anaerobic infections by MacDonald, Socransky & Gibbons (1963) showed that *B. melaninogenicus* was essential in all the mixtures tested. The ability of this organism to produce a collagenase has also encouraged the opinion that it may be involved in the destruction of periodontal tissues but although *B. melaninogenicus* may well be contributory, there is no conclusive evidence that this species is an essential component of the plaque which initiates periodontal disease.

A number of animal experiments on periodontal disease have been reported, and the value of rodents in particular for this work has been discussed by Jordan (1971). The species *Actinomyces viscosus*, originally described as *Odontomyces viscosus* (Howell *et al.*, 1965), has been shown by Jordan, Keyes & Bellack (1972) to induce periodontal disease in rodents. *A. naeslundii* can also be used to produce the disease in animals (Socransky, Hubersak & Propas, 1970). The frequent presence of these two *Actinomyces* species in dental plaque has been noted by several workers, and our own studies on plaque would confirm these observations.

In the last few years considerable interest has centred on the possibility that immunological mechanisms may play an important part in the aetiology of periodontal disease (Ivanyi & Lehner, 1970, 1971; Ivanyi, Wilton & Lehner, 1972; Ranney, 1970; Ranney & Zander, 1970; Baram & Arnold, 1970; Mergenhagen, Tempel & Snyderman, 1970). The precise nature of the immunological response to antigens in dental plaque and its significance in the pathogenesis of periodontal disease remains to be determined, and this is currently one of the major growth areas in dental research. Amongst the bacteria whose antigenic potential has been investigated and thought to be significant in immunological experiments are *Bacteroides*, *Actinomyces*, and *Veillonella* species.

The application of improved anaerobic techniques to the study of the flora associated with the gingival crevice and periodontal disease has been attempted by a few workers (Aranki *et al.*, 1969; Gordon *et al.*, 1971) and this development should help to increase our understanding of the situation. Two recent reports provide some evidence that anaerobic Gram-positive rods may be of particular significance in the aetiology of periodontal disease (Beveridge & Goldner, 1973; Sabiston & Grigsby, 1972).

ACUTE ULCERATIVE GINGIVITIS (A.U.G., Vincent's Gingivitis)

Acute ulcerative gingivitis is one of the more common acute inflammatory conditions affecting the mouth, and it is probably caused by anaerobic organisms. The patient usually presents with painful and bleeding gums, malaise and a foul taste in the mouth. Characteristically, the breath is foetid and there may be cervical lymphadenopathy but there is not usually a significant rise in temperature, and if fever is present the diagnosis is more likely to be herpetic gingivostomatitis. Typically the interproximal gingival papillae are ulcerated and may be covered with yellow-grey slough. The clinical features, differential diagnosis, epidemiology and other aspects of this disease have been reviewed by a number of authors, including Goldhaber & Giddon, (1964), Gordon, (1971), and Barnes, Bowles & Carter, (1973). In Europe and the U.S.A. the majority of cases are found amongst young adults, but in Nigeria the peak incidence is at a much lower age (Emslie, 1963; Sheiham, 1966).

Although AUG is thought to be due to a bacterial infection, it is not apparently communicable by normal contact. Outbreaks of the disease, such as the classical 'Trench Mouth' observed during war-time, are probably attributable to similar environmental conditions rather than transmission of the causative agent. Although bacterial debris from the gingivae of patients with AUG may produce suppurative lesions when injected into animals, it has not so far proved possible to reproduce the oral manifestations of the disease experimentally (Hampp and Mergenhagen, 1961, 1963; Rosebury, 1962).

Microscopic smears from the lesions in AUG typically show an abundance of spirochaetes and Gram-negative fusiform organisms, the so-called 'Fuso-Spirochaetal Complex', together with other morphological types. Although these organisms have for many years been regarded as the causative agents in the disease, a very similar microscopic appearance can sometimes be shown with material taken from the gingival region of subjects not suffering from AUG.

Electron microscopic examination of the gingival lesions in AUG has provided evidence of invasion by spirochaetes (Heylings, 1967; Listgarten, 1965), whereas no such penetration was observed in chronic gingivitis material (Freedman, Listgarten & Taichman, 1968). Listgarten (1965) described four main histological zones, all of which contained spirochaetes. The innermost zone showed high numbers of large spirochaetes and no other types of micro-organisms were observed. These studies indicate that the oral spirochaetes may possess a greater ability to penetrate gingival tissue than the other bacteria associated with AUG. In a subsequent investigation (Listgarten & Lewis, 1967) the distribution of morphological types of spirochaetes from AUG lesions was reported. Pooled scrapings from subjects with AUG showed that approximately 72 per cent of the observed spirochaetes were of the 'intermediate' type, 28 per cent were 'small', whilst only 3 per cent were classified as 'large'.

Unfortunately, most studies on the microbiology of AUG have relied exclusively upon microscopic evidence and few cultural investigations have been reported. Now that improved techniques for anaerobic bacteriology are available, the microbial aetiology of AUG might very usefully be re-examined.

Several forms of therapy have been recommended for treatment of this condition, ranging from local cleansing of the mouth to systemic antimicrobial chemotherapy. AUG usually occurs in patients with dirty mouths, and some attempt to clean up the mouth is invariably desirable. However, since the gingivae are usually painful, swollen and haemorrhagic, it may not be possible to institute vigorous oral hygiene measures until the acute phase has been controlled.

Penicillin is highly effective in the treatment of AUG and has been widely used, but since other equally effective agents are now available it is probably best to reserve it for other infections. The efficacy of metronidazole for this condition is now well established and numerous reports of successful clinical trials have appeared in recent years (Wade *et al.*, 1961; Davies, McFadzean & Squires, 1964; Shin, Squires & McFadzean, 1965; Duckworth *et al.*, 1966; Glenwright & Sidaway, 1966; Stephen *et al.*, 1966; Fletcher & Plant, 1966; Emslie, 1967, 1971; Proctor & Baker, 1971). Another similar compound, nitrimidazine, has recently been shown to be equally effective for the treatment of AUG (Lozdan *et al.*, 1971).

Metronidazole is known to be active only against anaerobic micro-organisms, and the effectiveness of this drug in the treatment of AUG provides supporting evidence for the idea that the disease is caused by anaerobes. The sensitivity of various anaerobic bacteria to metronidazole has been recorded by Tally, Sutter & Finegold, (1972) and Finegold & Rosenblatt, (1973), and is discussed in Chapter 4.

The action of metronidazole against oral spirochaetes *in vitro* has been demonstrated by Davies & Stirland (1970) and Leclair & Keetin (1970). Earlier studies (Fitzgerald & Hampp, 1952; Hampp & Fitzgerald, 1959) had shown that oral spirochaetes were susceptible to a number of other anti-microbial agents, including penicillin, erythromycin, bacitracin, streptomycin, and vancomycin. Since vancomycin has been shown by Emslie & Ashley, (1971) to be effective in treating patients with AUG, although Finegold and Rosenblatt, (1973) reported that this drug had poor or inconsistent activity against most of the anaerobes tested, this may provide a little more indirect evidence for the role of spirochaetes in the disease.

Takazoe *et al.*, (1973) found that experimental infections in guinea pigs, using mixtures of two *Bacteroides* strains together with an 'anaerobic *Corynebacterium*', could be completely prevented by administration of metronidazole.

CANCRUM ORIS

Cancrum oris or noma is a condition which results in massive soft tissue necrosis, often with involvement of underlying bone and subsequent sequestration. Although the aetiology is not completely understood, it appears that the disease always follows an attack of AUG which extends from the gingivae to the adjacent soft tissues. Cancrum oris is rarely seen today in Europe and the U.S.A., but is regularly encountered in some countries. A detailed report of the clinical features, predisposing factors and epidemiology of this condition in Nigeria has been published by Emslie (1963).

As with AUG, smears taken from the lesions in cancrum oris show an abundance of spirochaetes and fusiforms, together with other organisms. The limited cultural examinations which have been recorded indicate that *Bacteroides melaninogenicus* can be isolated from some cases. However, in Emslie's study the samples obtained for culture were transported from Nigeria to Boston before laboratory investigations were initiated, and it is probable that many of the anaerobes present originally did not survive the journey.

Analysis of possible predisposing factors indicated that prior infection with herpex simplex virus, measles, smallpox or malaria, and anaemia secondary to infestations with parasites might be significant. On the other hand, nutritional deficiencies, ABO blood groups, acatalasaemia and other systemic factors could not be incriminated as predisposing factors in the series of patients investigated (Emslie, 1963). Further epidemiological data on the prevalence of AUG and cancrum oris in Nigeria has been provided by Sheiham (1966).

It appears that this very unpleasant and disfiguring condition arises from a severe reaction to bacteria that are normally present in the mouth, particularly those of the gingival crevice region. It is highly probable that the main culprits are anaerobes and it would be interesting to carry out a more detailed microbiological investigation on this disease. It is possible that concurrent immunological studies might also prove to be rewarding.

ACTINOMYCOSIS

Actinomycosis of the head and neck region is thought to arise by infection of deeper tissues with *Actinomyces* originating from the oral commensal flora. Frequently the organisms are isolated in mixed culture from the pus, often in association with other anaerobic species. The extensive literature on commensal and pathogenic *Actinomyces* species in man has been reviewed quite recently (Bowden & Hardie, 1973) and will not be considered in detail here. In addition to the two serotypes of *A. israelii* (Holm, 1950), strains of *A. naeslundii* (Coleman, Ceorg & Rozzell, 1969) are also known to be causative agents in human actinomycosis. *Arachnia propionica* (Buchanan & Pine, 1962) was originally obtained from a case of lacrimal canaliculitis, and recently 11 further cases of actinomycosis caused by this organism have been reported (Brock *et al.*, 1973), including one neck abscess and a finger wound following a human bite. Although *Arachnia propionica* produces disease which is clinically indistinguishable from that due to *Actinomyces israelii*, the two organisms can be distinguished by cell wall analysis, gas-liquid chromatography of acid end products, and serology.

Actinomyces viscosus has only occasionally been reported from cases of actinomycosis, but this organism may play a role in periodontal disease as mentioned previously. The pathogenic potential of the other *Actinomyces* species found regularly in the mouth, *A. odontolyticus*, is not known.

Actinomyces (Bifidobacterium) eriksonii, originally isolated from a brain abscess (Georg *et al.*, 1965), has been observed in the tonsillar crypts, but does not seem to have been isolated regularly from the mouth, although one strain was found in root surface caries by Jordan and Hammond (1972). The animal pathogen, *A. bovis*, is not known to occur in humans, although many authors in the past have used this name incorrectly as a synonym for *A. israelii*.

OTHER ORAL INFECTIONS

Infections due to *Bacteroides* species have been reported with increasing regularity in the medical literature and numerous reviews and case reports have appeared recently (e.g. Felner & Dowell, 1970, 1971; Love, 1972; Mitchell, 1973; Nobles, 1973; Okubadejo, Green & Payne, 1973). Although the human mouth normally harbours several types of *Bacteroides*, there are few well-documented cases of local oral infections due to these organisms. Dormer & Babett (1972) have pointed out that *Bacteroides* may often be overlooked in oral specimens, and it is likely that a much higher isolation rate would be obtained by the use of improved anaerobic techniques. When *Bacteroides* strains are isolated from dental abscesses they are usually in mixed culture, often associated with *Actinomyces* species.

Two cases of osteomyelitis of the mandible due to *Bacteroides* have been recorded by Leake (1972). One of these was caused by *B. fragilis* in association with anaerobic streptococci, the other by a mixture of *B. melaninogenicus*, *Actinomyces israelii* and some unspecified 'diphtheroids'. Both cases responded successfully to treatment with clindamycin.

The possibility that bacteraemias of dental origin may include *Bacteroides* and other organisms is of particular interest to the dental surgeon in relation to prophylaxis for bacterial endocarditis in susceptible patients. Khairat (1966) stressed the importance of anaerobes in post-extraction bacteraemias. In his study 46/115 strains isolated from the blood after tooth extraction in 100 patients were anaerobes. These included 16 strains of *Bacteroides corrodens*, and various other *Bacteroides*, fusiforms and anaerobic cocci. The cases reported by Felner & Dowell (1971) showed that in the 9 patients with *B. oralis* bacteraemia there was a 70 per cent mortality, whereas none of the 4 patients with *B. melaninogenicus* died. The fact that the mouth may be the portal of entry for organism causing endocarditis is well documented and has recently been re-emphasized in a review of anaerobic infections by Finegold & Rosenblatt (1973). These authors also suggested that some of the cases of endocarditis from which no positive cultures have been obtained may be due to undetected anaerobes.

An interesting recent report by Sans & Crowder (1973) gives details of a case of subacute bacterial endocarditis due to *Eubacterium aerofaciens*. The source of infections in this patient may well have been a dental extraction which preceded the onset of endocarditis, although the time interval of two months makes it difficult to establish the connection with confidence.

Fortunately for the dental surgeon, wounds in the mouth become infected comparatively rarely. Infection of post-extraction wounds, leading to the painful condition known as 'dry socket', does occur occasionally, particularly following extraction of wisdom teeth. MacGregor (1970) has estimated that dry socket follows approximately 3 per cent of all tooth extractions. MacGregor & Hart (1969) found that staphylococci, streptococci and coryneforms were the organisms which could most regularly be isolated from post-extraction sockets, but a significant number also yielded anaerobes. No relationship was established between the flora of the socket immediately after extraction and the subsequent development of a dry socket.

Examination of Gram-stained smears from infected sockets showed the presence of a variety of morphological types, and fusiforms and spirochaetes appeared to be conspicuous before treatment was commenced (MacGregor, 1970). Detailed cultural studies on the anaerobic flora of dry socket might well be rewarding.

HALITOSIS

Foul-smelling breath is a fairly common complaint which may be caused by a variety of systemic and local conditions. Chronic periodontal disease, especially in advanced stages, is often accompanied by bad breath, and a particularly foetid odour is usually observed in patients suffering from acute ulcerative gingivitis. It is probable that the oral microflora is responsible for the production of malodorous compounds in the mouth, and it is likely that certain anaerobes are the chief offenders in this respect.

Tonzetich (1971) has studied the sulphur compounds in mouth air by gas chromatography and showed that 90 per cent of the sulphur content could be accounted for as hydrogen sulphide and methyl mercaptan, both of which have an objectionable putrid odour. Smaller amounts of dimethyl sulphide were also identified. These were shown to be the major volatile sulphur compounds produced *in vitro* by human dental plaque from cysteine, cystine and methionine (Tonzetich & Carpenter, 1971).

McNamara, Alexander & Lee (1972) investigated the role of micro-organisms in the production of oral malodour, and concluded that Gram-negative bacteria were the principal offenders. Under *in vitro* conditions cultures of *Leptotrichia*, *Fusobacterium*, *Veillonella* and *Bacteroides* species were all found to produce unpleasant smells, whereas several Gram-positive species tested did not share this attribute. It was also observed that availability of glucose tended to reduce the potential of bacteria to produce unpleasant odours.

Some recent reports from Japan (Mizuno *et al.*, 1971; Ueno, Ninomiya & Sazuki, 1971; quoted by Takazoe *et al.*, 1973) suggest that halitosis may be reduced by metronidazole. These observations tend to support the idea that in some cases foul breath is caused by the metabolism of anaerobic bacteria. The most obvious way of treating halitosis due to oral organisms is by treatment of periodontal disease and vigorous dental plaque control.

REFERENCES

Aranki, A., Syed, S.A., Kenney, S.B. & Freter, R. (1969). Isolation of anaerobic bacteria from human gingiva and mouse cecum by means of a simplified glove box procedure. *Applied Microbiology*, 17, 568.

Bahn, A.M. (1970). Microbial potential in the etiology of periodontal disease. *Journal of Periodontology*, 41, 617.

Bailit, H.L., Balwin, D.C. & Hunt, E.E., Jr. (1964). Increasing prevalence of gingival *Bacteroides melaninogenicus* with age in children. *Archives of oral Biology*, 9, 435.

Baram, P. & Arnold, L. (1970). Immunologic aspects of periodontal disease. *Journal of Periodontology*, 41, 617.

Barnes, G.P., Bowles, W.F. & Carter, H.G. (1973). Acute Necrotizing ulcerative gingivitis: A survey of 218 cases. *Journal of Periodontology*, 44, 35.

Berg, J.O. & Nord, C.E. (1972). Isolation of Peptococci and Peptostreptococci in developing human dental plaque by maintaining continuous anaerobiosis. *Acta Odontologica Scandinavica*, 30, 503.

Berg, J.D. & Nord, C.E. (1973). A method for isolation of anaerobic bacteria from endodontic specimens. *Scandinavian Journal of Dental Research*, 81, 163.

Beveridge, T.J. & Goldner, M. (1973). Statistical relationship between the presence of human subgingival anaerobic diphtheroids and periodontal disease. *Journal of Dental Research*, 52, 451.

Bowden, G.H. & Hardie, J.M. (1971). Anaerobic organisms from the human mouth. In *Isolation of Anaerobes*. Ed. Shapton, D.A. & Board, R.G. Academic Press, London.

Bowden, G.H. & Hardie, J.M. (1973). Commensal and pathogenic *Actinomyces* species in man. In *Actinomycetales: Characteristics and practical importance*. Ed. Sykes, G. & Skinner, F.A. Academic Press, London.

Bowden, G.H. & Hardie, J.M., & Slack, G.L. (1974). Microbial variations in approximal plaque. *Caries Research*. (In Press).

Brock, D.W., Georg, L.K., Brown, J.M. & Hicklin, M.D. (1973). Actinomycosis caused by *Arachnia propionica*. Report of 11 cases. *American Journal of Clinical Pathology*, 59, 66.

Brown, L.R., Merril, S.S. & Allen, R.E. (1970). Microbiologic study of intraoral wounds. *Journal of Oral Surgery*, 28, 89.

Buchanan, B.B. & Pine, L. (1962). Characterization of a propionic acid producing actinomycete, *Actinomyces propionicus*, sp. nov. *Journal of General Microbiology*, 28, 305.

Burnett, G.W. & Scherp, H.W. (1968). Oral Microbiology and Infectious Disease, 3rd edition. Williams & Wilkins, Baltimore.

Carlsson, J. (1971). Bacterial populations associated with the periodontium. In *The prevention of periodontal disease* Ed. Eastoe, J.E., Picton, D.C.A. & Alexander, A.G. Henry Kimpton London.

Carlsson, J. (1973). Simplified gas chromatographic procedure for identification of bacterial metabolic products. *Applied Microbiology*, 25, 287.

Coleman, R.M., Georg, L.K. & Rozzell, A.R. (1969). *Actinomyces naeslundii* as an agent of human actinomycosis. *Applied Microbiology*, 18, 420.

Cross, T. & Goodfellow, M. (1973). Taxonomy and Classification of the Actinomycetes. In *Actinomycetales: Characteristics and Practical Importance*. Ed. Sykes, G. & Skinner, F.A. Academic Press London.

Davies, A.H., McFadzean, J.A. & Squires, S. (1964). Treatment of Vincent's stomatitis with metronidazole. *British Medical Journal*, 1, 1149.

Davies, R.M. (1972). General ecology of the commensal micro-flora of the mouth. In *Host resistance to commensal bacteria. The response to dental plaque*. Ed. MacPhee, T. Edinburgh: Churchill Livingstone.

Davies, R.M., & Stirland, R.M. (1970). The in-vitro sensitivity of oral spirochaetes to metronidazole. *Journal of Periodontal Research*, 5, 183.

Dormer, B.J. & Babett, J.A. (1972). Orofacial infection due to bacteroides, a neglected pathogen. *Journal of Oral Surgery*, 30, 658.

Drasar, B.S. (1967). Cultivation of anaerobic intestinal bacteria. *Journal of Pathology and Bacteriology*, 94, 417.

Duckworth, R., Waterhouse, J.P., Britton, D.E.R., Nuki, K., Sheiham, A., Winter, R. & Blake, G.C. (1966). Acute ulcerative gingivitis. A double-blind controlled clinical trial of Metronidazole. *British Dental Journal*, 120, 599.

Dwyer, D.M. & Socransky, S.S. (1968). Predominant cultivable micro-organisms inhabiting periodontal pockets. *British Dental Journal*, 124, 560.

Emslie, R.D. (1963). Cancrum Oris. *Dental Practitioner*, 13, 481.

Emslie, R.D. (1967). Treatment of acute ulcerative gingivitis. *British Dental Journal*, 122, 307.

Emslie, R.D. (1971). Clinical trials of treatment for acute periodontal diseases. *International Dental Journal*, 21, 33.

Emslie, R.D. & Ashley, F.P. (1971). Topical treatment of acute ulcerative gingivitis: A comparison of vancomycin with penicillin and metronidazole. *Paradontologie*, 25, 3.

Eskow, R.N. & Loesche, W.J. (1971). Oxygen tensions in the human oral cavity. *Archives of Oral Biology*, 16, 1127.
Feldmann, G. & Larje, O. (1966). The bacterial flora of submucous abscesses originating from chronic exacerbating osteitis. *Acta Odontologica Scandinavica*, 24, 129.
Fein, S. & Mohnac, A.M. (1973). Ludwig's angina infection: report of a case and associated systemic complications. *Journal of Oral Surgery*, 31, 785.
Felner, J.M. & Dowell, V.R. (1970). Anaerobic bacterial endocarditis. *The New England Journal of Medicine*, 283, 1188.
Felner, J.M. & Dowell, V.R. (1971). 'Bacteroides' bacteremia. *The American Journal of Medicine*, 50, 787.
Finegold, S.M., Marsh, V.H. & Bartlett, J.G. (1971). Anaerobic infections in the compromised host. In *Proceeding of International conference on Nosocomical Infections*, Ed. Brachman, P.S. & Eickhoff, T.C. American Hospital Association, Chicago, p.123.
Finegold, S.M. & Rosenblatt, J.E. (1973). Practical aspects of anaerobic sepsis. *Medicine*, 52, 311.
Fitzgerald, R.J. & Hampp, E.G. (1952). Inhibition of oral spirochaetes by antibiotic agents in vitro. *Journal of Dental Research*, 31, 20.
Fletcher, J.P. & Plant, C.G. (1966). As assessment of metronidazole in the treatment of AUG. *Oral Surgery*, 22, 729.
Freedman, H.L., Listgarten, M.A. & Taichman, N.S. (1968). Electron microscopic features of chronically inflamed human gingiva. *Journal of Periodontal Research*, 3, 313.
Fulghum, R.S., Wiggins, C.B. & Mullaney, T.P. (1973). Pilot study for detecting obligate anaerobic bacteria in necrotic dental pulps. *Journal of Dental Research*, 52, 637.
Genco, R.J., Evans, R.T. & Ellison, S.A. (1969). Dental research in microbiology with emphasis on periodontal disease. *Journal of the American Dental Association*, 78, 1016.
Georg, L.K., Robertstad, G.W, Brinkman, S.A. & Hicklin, M.D. (1965). A new pathogenic anaerobic *Actinomyces* species. *Journal of Infectious Diseases*, 115, 88.
Gibbons, R.J., Socransky, S.S., Sawyer, S., Kapsimalis, B. & MacDonald, J.B. (1963). The microbiota of the gingival crevice area of man. II. The predominant cultivable organisms. *Archives of Oral Biology*, 8, 281.
Glenwright, H.D. & Sidaway, A. (1966). The use of metronidazole in the treatment of AUG. *British Dental Journal*, 121, 174.

Goldhaber, P. & Giddon, D. (1964). Present concepts concerning the aetiology and treatment of AUG. *International Dental Journal*, 14, 468.

Gordon, D.F., Stutman, M. & Loesche, W.J.(1971). Improved isolation of anaerobic bacteria from the gingival crevice area of man. *Applied Microbiology*, 21, 1046.

Gordon, L. (1971). Current status of ANUG. *Oral Health*, 61, 23.

Greene, J.C. (1963). Oral hygiene and periodontal disease. *American Journal of Public Health*, 53, 913.

Hadi, A.W., & Russell, C. (1968). Quantitative estimations of fusiforms in saliva from normal individuals and cases of acute ulcerative gingivitis. *Archives of Oral Biology*, 13, 1371.

Hadi, A.W. & Russell, C. (1969). Fusiforms in gingival material. *British Dental Journal*, 126, 82.

Hampp, E.G. (1947). Bacteriologic investigations of the oral spirochetal flora in ulcerative stomatitis (Vincent's infection). *American Journal of Orthodontics and Oral Surgery*, 33, 666.

Hampp, E.G. (1957). Isolation and identification of spirochetes obtained from unexposed canals of pulp-involved teeth. *Oral Surgery*, 10, 1100.

Hampp, E.G. & Fitzgerald, R.J. (1959). The in vitro effects of antibiotics on oral spirochaetes. *Journal of Dental Research*, 38, 947.

Hampp, E.G. & Mergenhagen, S.E. (1961). Experimental infections with oral spirochaetes. *Journal of Infectious Diseases*, 109, 43.

Hampp, E.G. & Mergenhagen, S.E. (1963). Experimental intracutaneous fusobacterial and fusospirochaetal infections. *Journal of Infectious Diseases*, 112, 84.

Hardie, J.M. & Bowden, G.H. (1974), The normal microbial flora of the mouth. In *The normal microbial flora of man*,', *Society of Applied Bacteriology Symposium Series No.3*. Academic Press (in press).

Heylings, R.T. (1967). Electron microscopy of acute ulcerative gingivitis (Vincent's type). Demonstration of the fusospirochaetal complex of bacteria within pre-necrotic gingival epithelium. *British Dental Journal*, 122, 51.

Holdeman, L.V. & Moore, W.E.C. (1972). Anaerobe Laboratory Manual Blacksburg, Virginia: The Virginia Polytechnic Institute and State University Anaerobe Laboratory.

Holm, P. (1950). Studies on the aetiology of human actinomycosis. *Acta Pathologica et Microbiologica Scandinavica*, 27, 736.

Holmberg, K. & Hallander, H.O. (1973). Numerical taxonomy and laboratory identification of *Bacterionema matruchotii, Rothia dentocariosa, Actinomyces naeslundii, Actinomyces viscosus* and some related bacteria. *Journal of General Microbiology,* 76, 43.

Howell, A. Jr., Jordan, H.V., Georg, L.K. & Pine, L. (1965). *Odontomyces viscosus,* gen. nov., spec. nov., a filamentous micro-organism isolated from periodontal plaque in hamsters. *Sabouraudia,* 4, 65.

Hungate, R.E. (1966). The Rumen and its Microbes. Academic . Press: New York.

Ivanyi, L. & Lehner, T. (1970). Stimulation of lymphocyte transformation by bacterial antigens in patients with periodontal disease. *Archives of Oral Biology,* 15, 1089.

Ivanyi, L. & Lehner, T. (1971). Lymphocyte transformation by sonicates of dental plaque in human periodontal disease. *Archives of Oral Biology,* 16, 1117.

Ivanyi, L., Wilton, J.M.A. & Lehner, T. (1972). Cell-mediated immunity in periodontal disease; cytotoxicity, migration inhibition and lymphocyte transformation studies. *Immunology,* 22, 141.

Jordan, H.V. (1971). Rodent model systems in periodontal disease research. *Journal of Dental Research,* 50, 236.

Jordan, H.V. & Hammond, B.F. (1972). Filamentous bacteria isolated from human root surface caries. *Archives of Oral Biology,* 17, 1333.

Jordan, H.V., Keyes, P.H. & Bellack, S. (1972). Periodontal lesions in hamsters and gnotobiotic rats infected with *Actinomyces* of human origin. *Journal of Periodontal Research,* 7, 21.

Jordan, H.V. & Sumney, D.L. (1973). Root surface caries: Review of the literature and significance of the problem. *The Journal of Periodontology,* 44, 158.

Khairat, O. (1966). An effective antibiotic cover for the prevention of endocarditis following dental and other post-operative bacteraemias. *Journal of Clinical Pathology,* 19, 561.

Kelstrup, J. (1966). The incidence of *Bacteroides melanino-genicus* in human gingival sulci, and its prevalence in the oral cavity at different ages. *Periodontics,* 4, 14.

Kenney, E.B. & Ash, M.M. (1969). Oxidation-reduction potential of developing plaque, periodontal pockets and gingival sulci. *Journal of Periodontology,* 40, 630.

Killey, H.C., Seward, G.R. & Kay, L.W. (1971). An Outline of Oral Surgery, Parts I and II. Bristol: Wright.

Killgore, G.E., Starr, S.E., Bene, V.E.D., Whaley, D.N. & Dowell, V.R. (1973). Comparison of three anaerobic systems for the isolation of anaerobic bacteria from clinical specimens. *American Journal of Clinical Pathology,* 59, 552.

Klainer, A.S. & Beisel, W.R. (1969). Opportunist Infection: A review. *American Journal of Medical Science*, 258, 431.
Leake, D.L. (1972). Bacteroides osteomyelitis of the mandible. A report of two cases. *Oral Surgery*, 34, 585.
Leclair, R.A. & Keetin, L.M. (1970). In vitro susceptibility of oral spirochaetes to metronidazole. *Journal of Dental Research*, 49, 1513.
Levison, M.E. (1973). The importance of anaerobic bacteria in infectious diseases. *Medical Clinics of North America*, 57, 1015.
Listgarten, M.A. (1965). Electron microscopic observations on the bacterial flora of acute necrotizing ulcerative gingivitis. *Journal of Periodontology*, 36, 328.
Listgarten, M.A. & Lewis, D.W. (1967). The distribution of spirochaetes in the lesion of acute necrotizing ulcerative gingivitis: an electron microscope and statistical survey. *Journal of Periodontology*, 38, 379.
Listgarten, M.A. & Socransky, S.S. (1965). Electron microscopy as an aid in the taxonomic differentiation of oral spirochaetes. *Archives of Oral Biology*, 10, 127.
Loe, H., Theilade, E. & Jenson, S.B. (1965). Experimental gingivitis in man. *Journal of Periodontology*, 36, 177.
Loesche, W.J. (1968). Importance of nutrition in gingival crevice microbial ecology. *Periodontics*, 6, 245.
Loesche, W.J. (1969). Oxygen sensitivity of various anaerobic bacteria. *Applied Microbiology*, 18, 723,
Loesche, W.J. & Gibbons, R.J. (1965). A practical scheme for identification of the most numerous oral Gram-negative anaerobic rods. *Archives of Oral Biology*, 10, 723.
Loesche, W.J., Gibbons, R.J. & Socransky, S.S. (1965). Biochemical characteristics of *Vibrio sputorum* and relationship to *Vibrio bubulus* and *Vibrio fetus*. *Journal of Bacteriology*, 89, 1109.
Loesche, W.J., Hockett, R.N. & Syed, S.A. (1972). The predominant cultivable flora of tooth surface plaque removed from institutionalized subjects. *Archives of Oral Biology*, 17, 1311.
Loesche, W.J. & Syed, S.A. (1973). The predominant cultivable flora of carious plaque and carious dentine. *Caries Research*, 7, 201.
Love, W.C. (1972). Clinical aspects of Bacteroides infections. In *Host resistance to commensal bacteria. The response to dental plaque*. Ed. MacPhee, T. Churchill Livingstone, Edinburgh, p.298.
Lozdan, J., Sheiham, A., Pearlman, B.A., Keizer, B., Rachanis, C.C. & Meyer, R. (1971). The use of nitrimidazine in the treatment of AUG. *British Dental Journal*, 130, 294.

MacDonald, J.B., Socransky, S.S. & Gibbons, R.J. (1963). Aspects of the pathogenesis of mixed anaerobic infections of mucous membranes. *Journal of Dental Research*, 42, 529.

MacGregor, A.J. (1970). The bacteriology of dry socket; the effect of zinc oxide and oil of cloves. In *Oral Surgery: Transactions of the Third International Conference on Oral Surgery*. Ed. Walker, R.V. Livingstone: Edinburgh.

MacGregor, A.J. & Hart, P. (1969). Effect of bacteria and other factors on pain and swelling after removal of ectopic mendibular third molars. *Journal of Oral Surgery*, 27, 174.

McCarthy, C., Snyder, M.L. & Parker, R.B. (1965). The indigenous oral flora of man - I. The newborn to the 1-year old infant. *Archives of Oral Biology*, 10, 61.

McNamara, T.F., Alexander, J.F. & Mary Lee. (1972). The role of micro-organisms in the production of oral malodor. *Oral Surgery*, 34, 41.

Mergenhagen, S.E., Tempel, T.R. & Snyderman, R. (1970). Immunologic reactions and periodontal inflammation. *Journal of Dental Research*, 49, 256.

Mikx, F.H.M., Hoeven, J.S. van der., Konig, K.G., Plasschaert, A.J.M. & Guggenheim, B. (1972). Establishment of defined microbial ecosystems in germ free rats. 1. The effect of the interaction of *Streptococcus mutans* or *Streptococcus sanguis* with *Veillonella alkalescens* on plaque formation and caries activity. *Caries Research*, 6, 211.

Mitchell, A.A.B. (1973). Incidence and isolation of Bacteroides species from clinical material and their sensitivity to antibiotics. *Journal of Clinical Pathology*, 26, 738.

Mizuno, K., Abe, T., Kawai, M., Shiraiwa, A. & Akita, Y. (1971). Effects of metronidazole (Flagyl) on halitosis in periodontal disease. *Aichi-Gakuin Shigaku-Zaschi*, 8, 192 (in Japanese).

Möller, A.J.R. (1966). *Microbiological Examination of Root Canals and Periapical Tissues of Human Teeth.* Methodological studies. Thesis, distributed by Akademiforglaget, Goteborg.

Moore, J.R. & Russell, C. (1972). Bacteriological investigation of dental abscesses. *Dental Practitioner*, 22, 390.

Moore, W.E.C., Cato, E.P. & Holdeman, L.V. (1969). Anaerobic bacteria of the gastrointestinal flora and their occurrence in clinical infections. *Journal of Infectious Diseases*, 119, 641.

Nobles, E.R. (1973). Bacteroides infections. *Annals of Surgery*, 177, 601.

Nolte, W.A. (1973). Oral Microbiology. 2nd Edition. St. Louis: The C.V. Mosby Company. London: Henry Kimpton .

Okubadejo, O.A., Green, P.J. & Payne, D.J.H. (1973). Bacteroides infections among hospital patients. *British Medical Journal*, 2, 212.
Pine, L. (1970). Classification and phylogenetic relationship of microaerophilic actinomycetes. *International Journal of Systematic Bacteriology*, 20, 445.
Proctor, D.B. & Baker, C.G. (1971). Treatment of acute necrotizing ulcerative gingivitis with metronidazole. *Journal of the Canadian Dental Association*, 37, 376.
Ranney, R.R. (1970). Specific antibody in gingiva and submandibular nodes of monkeys with allergic periodontal disease. *Journal of Periodontal Research*, 5, 1.
Ranney, R.R. & Zander, H.A. (1970). Allergic peridontal disease in sensitized squirrel monkeys. *Journal of Periodontology*, 41, 12.
Ritz, H.L. (1967). Microbial population shifts in developing human plaque. *Archives of Oral Biology*, 12, 1561.
Rosebury, T. (1962). Micro-organisms indigenous to man. McGraw Hill, New York.
Russell, A.L. (1963). International Nutrition Surveys; A summary of preliminary findings. *Journal of Dental Research*, 42, 233.
Sabiston, C.B. & Grigsby, W.R. (1972). Anaerobic bacteria from the advanced periodontal lesion. *Journal of Periodontology*, 43, 199.
Sans, M.D. & Crowder, J.G. (1973). Subacute bacterial endocarditis caused by *Eubacterium aerofaciens*. *American Journal of Clinical Pathology*, 59, 576.
Sheiham, A. (1966). An epidemiological survey of acute ulcerative gingivitis in Nigerians. *Archives of Oral Biology*, 11, 937.
Sheiham, A. (1970). Dental cleanliness and chronic periodontal disease. Studies on populations in Britain. *British Dental Journal*, 129, 413.
Shin, D.L.D., Squires, S. & McFadzean, J.A. (1965). The treatment of Vincent's disease with Metronidazole. *Dental Practitioner*, 15, 275.
Slack, G.L. & Bowden, G.H. (1965). Preliminary studies of experimental dental plaque *in vivo*. *Advances in Fluorine Research and Dental Caries Prevention*, 3, 193.
Socransky, S.S. (1970). Relationship of bacteria to the aetiology of periodontal disease. *Journal of Dental Research*, 49, 203.
Socransky, S.S., Gibbons, R.J., Dale, A.C., Bortnick, E., Rosenthal, E. & MacDonald, J.B. (1963). The microbiota of the gingival crevice area of man. Total microscopic and viable counts of specific organisms. *Archives of Oral Biology*, 8, 275.

Socransky, S.S., Listgarten, M.A., Hubersak, C., Cormore, J. & Clark, A. (1969). Morphological and biochemical differentiation of three types of small oral spirochaetes. *Journal of Bacteriology*, 98, 878.

Socransky, S.S., Hubersak, C. & Propas, D. (1970). Induction of periodontal destruction in gnotobiotic rats by a human oral strain of *Actinomyces naeslundii*. *Archives of Oral Biology*, 15, 993.

Socransky, S.S. & Manganiello, S.D. (1971). The oral microbiota of man from birth to senility. *The Journal of Periodontology*, 42, 485.

Stephen, K.W., McLatchie, M.F., Mason, D.K., Noble, H.W. & Stevenson, D.M. (1966). Treatment of acute ulcerative gingivitis (Vincent's type). *British Dental Journal*, 121, 313.

Sutter, V.L., Attebery, H.R., Rosenblatt, J.E., Bricknell, K.S. & Finegold, S.M. (1972). *Anaerobic Bacteriology Manual*. School of Medicine, University of California.

Takazoe, L., Okuda, K., Yamamoto, A. & Nakamura, T. (1973). The effect of Metronidazole on the experimental mixed infection of oral anaerobes. *Bulletin of Tokyo Dental College*, 14, 1.

Tally, F.P., Sutter, V.L. & Finegold, S.M. (1972). Metronidazole versus anaerobes. In vitro data and initial clinical observations. *California Medicine*, 117, 22.

Theilade, E., Wright, W.H., Jensen, S.B. & Loe, H. (1966). Experimental gingivitis in man. II. A longitudinal clinical and bacteriological investigation. *Journal of Periodontal Research*, 1, 1.

Tonzetich, J. (1971). Direct gas chromatographic analysis of sulphur compounds in mouth air in man. *Archives of Oral Biology*, 16, 587.

Tonzetich, J. & Carpenter, P.A.W. (1971). Production of volatile sulphur compounds from cysteine, cystine and methionine by human dental plaque. *Archives of Oral Biology*, 16, 599.

Ueno, K., Ninomiya, K. & Sazuki, S. (1971). Effects of Flagyl (Metronidazole) on Fetor Oris in periodontal disease. *Shikwa Gakuho*, 70, 1465 (in Japanese).

Van Reenan, J.F. & Coogan, M.M. (1970). Clostridia isolated from human mouths. *Archives of Oral Biology*, 15, 845.

Veron, M. & Chatelain, R. (1973). Taxonomic study of the genus *Campylobacter* Sebald and Veron and designation of the neotype strain for the Type Species, *Campylobacter fetus* (Smith & Taylor) Sebald & Vernon. *International Journal of Systematic Bacteriology*, 23, 122.

Wade, A.B., Blake, G.C., Manson, J.D., Anderson, E. & Arnold, M. (1961). Comparative assessment of three medicaments for the treatment of acute ulcerative gingivitis of the Vincent's type. *British Dental Journal*, 111, 280.
Williams, R.E.O. (1973). Benefit and mischief from commensal bacteria. *Journal of Clinical Pathology*, 26, 811.
Zabransky, R.J. (1970). Isolation of anaerobic bacteria from clinical specimens. *Mayo Clinic Proceedings*, 45, 256.

DISCUSSION

DR DOWSETT (BASILDON): In view of what you've been saying about these organisms is benzylpenicillin going to be the best prophylactic treatment for people likely to have valve disease?
MR HARDIE: I wouldn't like to give a definitive answer to that partly because in my own studies I don't usually examine sensitivity so I can't give data from my own experience. I think I would like to circumvent this by saying that there are organisms in the mouth which are not going to be affected much by benzylpenicillin, but if you are going to play the odds, the organisms which seem to get into the blood in the highest numbers usually are sensitive. There are bound to be small numbers of organisms that are going to be resistant.
PROF. GARROD: I would just like to say that I'm not in the least surprised in Mr Hardie's findings or somebody elses that he presented that *Bacteroides fragilis* does not figure very much in the mouth. I have been accustomed to regard the source of *Bacteroides fragilis* infection as being somewhere below the waist, but I would like to know whether other people agree with that. I don't know whether I may be permitted to add this, Sir, but somebody has said today that *Bacteroides fragilis* infections are the most important among those we are considering. I'm not at all sure that's true. I don't think they're the most virulent of them and I'm not sure that they're the commonest. I think this impression has got around because *Bacteroides fragilis* is of all the Gram-negative anaerobes quite the easiest to grow and in some peoples hands it's the only one they ever do grow!
MR HARDIE: Could I just comment and thank you for those remarks. What surprises me is that if we can grow all the others we should be able to grow *Bacteroides fragilis* and yet we don't seem to find it.

7. Head and Neck, Bone and Joint, and Chest Infections, Bacteraemia and Endocarditis due to Non-sporeforming Anaerobes

J. G. BARTLETT

CENTRAL NERVOUS SYSTEM INFECTIONS

Pyogenic infections of the central nervous system that commonly involve anaerobic bacteria are cerebral abscess, subdural empyema, and septic thrombophlebitis of the cortical veins and venous sinuses. Initiating foci in these cases are often infections of adjacent sites such as mastoiditis, otitis media or sinusitis. Intracranial extension is facilitated by the propensity of anaerobes to invade tissue and blood vessels. An additional common underlying lesion with brain abscesses is hematogenous dissemination from a distant site, particularly chronic suppurative infections of the lung (Cohen, 1934).

Pyogenic meningitis is seldom caused by oxygen-sensitive bacteria. Among 207 cases studied at Massachusetts General Hospital, there was only a single anaerobic isolate (Swartz & Karchmer, 1974). The recovery of these microbes from cerebrospinal fluid should alert the physician to the possibility of an unusual underlying lesion such as communicating brain abscess or otitis media.

Of the central nervous system infections likely to involve anaerobes, the most complete bacteriological data are available for brain abscesses. In 787 cases reported in 15 series, anaerobes were recovered in 12 per cent while sterile abscesses were noted in 22 per cent (Swartz & Karchmer, 1974). This type of analysis fails to take into account certain established bacterial associations with different patient populations as well as the marked variations in laboratory techniques. The latter point is illustrated by the frequent reports of 'sterile' cultures, which undoubtedly reflect the failure of many laboratories to use adequate anaerobic methods. Indeed, in some reports there was no attempt to recover oxygen-sensitive forms. Restricting analysis to studies in which at least some form of anaerobic cultures were routinely performed, the recovery rate of anaerobes has generally been 23-35 per cent (Table 7.1) (McFarlan, 1943; Beerens & Tahon-Castel, 1965; Swartz & Karchmer, 1974). These reports were non-selective in terms of the patients studied and the incidence of sterile cultures was only 2-9 per cent.

Table 7.1. Incidence of Anaerobes in Brain Abscess.

Literature source	Cases Studied	Cases with anaerobic isolates	Cases with sterile cultures
Swartz & Karchmar (In Press)	70	16 (23%)	6 (9%)
Beerens & Tahon-Castel (1965)	37	11 (30%)	2 (5%)
McFarlan (1943)	48	17 (35%)	1 (2%)
Heineman & Braude (1963)	18	16 (89%)	2 (11%)

The highest yield of anaerobes in brain abscesses has been reported by Heineman & Braude (1963), who studied 18 cases with optimal bacteriological methods. Anaerobic bacteria were recovered in 16 cases (89 per cent) and these organisms were the exclusive isolates in 10 (56 per cent). The predominant anaerobes were peptostreptococci (11 cases), *Bacteroides (5)*, fusobacteria (4), and *Actinomyces (3)*. The nature of the predisposing conditions in this study may to some extent account for the high yield of anaerobes; most of the patients had otitis or sinusitis as the initiating focus of infection. In this setting, as illustrated by these investigators, anaerobes are almost universally responsible.

Metastatic brain abscesses are also likely to involve anaerobic bacteria. Gregory, Messner & Zinneman (1967) reported 13 cases and noted that anaerobes were recovered in five of the six instances in which anaerobic cultures were performed. Similarly, brain abscess complicating cyanotic heart disease is often an anaerobic infection. A review of the reported experience with such cases indicated that oxygen-sensitive bacteria were isolated in 10 of the 32 culture-positive cases (Newton, 1956).

Brain abscess resulting from head trauma or neurosurgery usually involves a different group of pathogens, and anaerobes have seldom been implicated. Brain abscess occurring in the immunologically compromised host is another setting in which these organisms appear to be relatively uncommon. Among 39 cases occurring in cancer patients studied at the Memorial

Sloan-Kettering Cancer Center, only three yielded anaerobic bacteria (Chernik, Armstrong & Posner, 1973).

The most important therapeutic manoeuvre in central nervous system suppurative collections is adequate surgical drainage. Occasionally, 'cerebritis' may be successfully treated with aggressive and early chemotherapy before an abscess forms (Heineman & Braude, 1963). Such infections, however, constitute a small portion of all cases and surgery is usually required.

We consider that chloramphenicol is the antibiotic of choice for anaerobic pyogenic infections of the central nervous system. This is based on the excellent *in vitro* activity of this agent against virtually all anaerobic bacteria (Martin, Gardner & Washington, 1972) as well as its excellent central nervous system penetration (Garrod & O'Grady, 1972). The serious nature of these infections easily justifies the rare haematological risks associated with chloramphenicol. Penicillin is an alternative antibiotic, providing that *B. fragilis* can be excluded. The actual incidence of this organism in brain abscess has not been defined since in most studies, the *Bacteroides* isolates have not been speciated. Clindamycin should not be used because the available information suggests that central nervous system penetration with this agent is relatively poor (Geddes, Sleet & Murdoch, 1964). Cephalosporins are inactivated in crossing the blood-brain barrier giving poor levels of the active compound, and should also be avoided (Mangi *et al.*, 1973).

INFECTIONS OF THE UPPER RESPIRATORY TRACT

Infections of the upper respiratory tract were once a major source of severe anaerobic sepsis. Gunn (1956) reviewed the literature on anaerobic Gram-negative bacillaemia reported prior to 1956. The oropharynx was the most common portal of entry, accounting for 54 of 173 cases (31 per cent). Many of these patients had a characteristic syndrome consisting of tonsillitis or pharangitis with jugular vein thrombophlebitis and disseminated septic emboli to the lungs or joints (Lemierre, 1936).

These infections are now infrequently encountered. This may be related to the widespread use of chemotherapy which will interrupt the natural course of events and thus mask late clinical features which would conform to the classical description. In more recent studies of anaerobic Gram-negative bacillaemia the oropharynx has been the suspected portal of entry in only 0-9 per cent of cases (Wilson *et al.*, 1972; Chow & Guze, 1974; Felner & Dowell, 1971; Nobles, 1973).

The most common anaerobic infections of the upper respiratory passages encountered in recent years are periodontal infections, chronic sinusitis, peritonsillar abscess, chronic otitis media, and mastoiditis. The bacteriology of these infections reflects the indigenous flora of the adjacent mucosal surfaces with the predominant organisms being anaerobic and microaerophilic streptococci, *Fusobacterium* and *Bacteroides melaninogenicus*. Studies from the Forsyth Dental Infirmary indicate that *Bacteroides melaninogenicus* may be particularly important. This organism proved to be an indispensable component of gingivitis exudate for transferable pathogenicity in experimental animals (MacDonald, Socransky & Gibbons, 1963). *B. melaniogenicus*, however, may be easily escape detection owing to its relatively fastidious growth requirements.

Precise data on the frequency of anaerobic infection in most of these conditions is not available. Many cases are treated empirically without the benefit of culture. Even when cultures are performed, anaerobic methods have seldom been employed. These problems are illustrated well with acute and chronic otitis media. There are several reports of the bacteriology of middle ear fluid obtained by needle aspiration in cases of acute otitis media (Feingold *et al.*, 1966; Coffey, 1966; Howie, Ploussard & Lester, 1970). Unfortunately anaerobic cultures have not been routinely performed in any large study. The failure to recover aerobic pathogens in 20-30 per cent of specimens suggests that oxygen-sensitive forms may be involved in a significant proportion of the cases. With chronic otitis the evidence for a prominent role of anaerobic bacteria is more compelling, although still indirect. Foetid discharge is a common feature of such infections and suppurative central nervous system complications of chronic otitis (and mastoiditis) usually involve anaerobic bacteria (Heineman & Braude, 1963). Again adequate bacteriological studies have not been performed and most culturally confirmed cases particularly those with unusual complications, appear in the form of case reports.

An additional factor in the paucity of good bacteriological studies of anaerobic upper respiratory tract infections relates to the difficulty in obtaining reliable specimens. For example, transnasally collected sinusitis fluid, exudate from the external ear canal in cases of otorrhea, gingivitis exudate, and throat swabs are quite meaningless for anaerobic culture unless special precautions are taken to avoid the indigenous flora of mucocutaneous surfaces.

Despite the paucity of information in most disease categories there are excellent data from a recent study of the bacteriology of chronic sinusitis (Frederick & Braude, 1974). Optimal bacteriological techniques were applied to sinusitis fluid which was carefully obtained at surgical drainage procedures. Among 83 patients, anaerobes were recovered from 43 (52 per cent) and these microbes were the exclusive isolates in 26 (31 per cent). The most common bacteria recovered in heavy growth were peptostreptococci, fusobacteria and *Bacteroides*. It is of interest that there was only a single instance in which only aerobic bacteria were recovered in large numbers.

Three less common anaerobic infections are related to the upper respiratory passages. Vincent's angina is an ulcerating membraneous tonsillar infection in which the foetid odour and Gram stain showing the characteristic forms of anaerobic bacteria, serve to differentiate this infection from other forms of exudative tonsillitis. Ludwig's angina is an acute, gangrenous, indurated, bilateral cellutitis of both sublingual and submaxillary spaces. The infection spreads rapidly along tissue planes without abscess formation or lymph node involvement. There may be acute respiratory tract obstruction necessitating emergency tracheostomy or intubation (Williams & Guralnick, 1943). Cervicofacial actinomycosis is an indurated ('woody hard') connective tissue infection often associated with draining sinuses or fistulae. This infection may involve any species of *Actinomyces*, almost invariably in association with other bacteria as well. *Actinomyces* are present in the normal oral flora and the infection is usually preceeded by dental manipulation, trauma or malignancy.

Benzylpenicillin is generally regarded as the agent of choice for anaerobic infections of the upper respiratory tract. This is based on the premise that *B. fragilis* is not a normal inhabitant of the oropharynx (Rosebury, 1962) and, consequently, is seldom involved in these infections. Ampicillin shows *in vitro* activity against anaerobes comparable to that of benzylpenicillin, while other penicillins (such as penicillin V and isoxazole penicillins) are distinctly less active (Finegold *et al.*, 1965; Finegold, Sugihara & Miller, 1967). Clindamycin and chloramphenicol are alternative agents for penicillin-allergic patients. Cephalosporins have an anaerobic spectrum comparable to penicillin, but are only about 25 per cent as active against most strains on a weight basis (Martin *et al.*, 1972). Tetracycline is erratic in its activity against anaerobic bacteria and should be avoided unless susceptibility tests for a specific isolate indicate acceptable activity (Martin *et al.*, 1972). Erythromycin, metronidazole, sulphonamides plus trimethoprim, and minocycline may be useful although there is little published clinical experience with these agents in anaerobic infections.

BONE AND JOINT INFECTIONS

Anaerobic bacteria are infrequent pathogens in suppurative arthritis. Among 78 cases studied at the Mayo Clinic these organisms were recovered from only two patients (Kelly, Wilkowske & Washington, 1973). Although culture techniques were not described, likely pathogens, usually *Staphylococcus aureus*, were recovered in every case. This mitigates against the theoretical contention that bacteriological methods were suboptimal.

Most studies of anaerobic pyogenic arthritis appear in the form of case reports. In 1969 Ziment reviewed the world's literature and added an additional case involving *B. fragilis* following intraarticular administration of steroids to the knee (Ziment, Davis & Finegold, 1969). There were 48 cases, most of which were observed in the pre chemotherapy era. At that time there appeared many reports of otorhinopharyngeal infection associated with *Fusobacterium necrophorum* bacteraemia and septic arthritis involving multiple joints. This type of infection has seldom been noted recently. In the more recent studies, the predominant isolates have been anaerobic cocci, *B. fragilis* and *B. melaninogenicus*. By contrast with anaerobic osteomyelitis, only 5 cases of pyogenic arthritis involved a mixed aerobic-anaerobic flora, and only one patient had more than one anaerobic species. Thus, a single anaerobe was recovered in pure culture in 42 of the 48 cases. The authors called attention to poor host resistance, underlying joint disease, and corticosteroid administration as important predisposing factors.

Therapy of anaerobic purulent arthritis should include appropriate drainage, immobilization and antimicrobial therapy. Benzylpenicillin is the agent of choice for infections not involving *B. fragilis*. The latter may be treated with clindamycin which also serves as an alternative to penicillin in other cases. Tetracycline and erythromycin may be useful in selected patients. Antimicrobial therapy should be given systemically; intra-articular administration is seldom necessary.

As with joint infections, anaerobes are seldom encountered in osteomyelitis. For example, Kelly *et al.*, (1970), recovered these organisms in only 7 of 82 (8 per cent) patients with osteomyelitis of the femur or tibia. A similar experience has been noted by other observers studying other forms osteomyelitis as well (Waldvogel, Medoff & Swartz, 1970).

The largest report of anaerobic osteomyelitis consists of 17 patients reported by Ziment, Miller & Finegold (1968). These infections were usually polymicrobial, there being an average of 5.4 isolates per case. This included 2.8 aerobic or facultative bacteria and 2.6 anaerobes. The predominant anaerobes, in order of prevalence, were *Bacteroides melaninogenicus*, *B. fragilis*, fusobacteria, peptococci and peptostreptococci.

In this study the primary sites of involvement were the foot and skull. Osteomyelitis usually resulted from spread of adjacent soft tissue infections. Osteomyelitis of the foot in diabetic patients was particularly common, accounting for nine of the 17 cases. Clues to the presence of anaerobes were foul discharge (14 cases), the distinctive morphological features of anaerobic bacteria on direct Gram stain (12), and the presence of gas in the soft tissues (4).

Therapy of osteomyelitis involving anaerobic bacteria is similar to that required by other forms of osteomyelitis. The most important factor in chronic cases is the meticulous surgical removal of all dead bone, surrounding scar, and granulation tissue. The involved bone should be immobilized and prolonged antimicrobial therapy is usually required. Drug selection should be based on bacteriological findings according to the previously noted recommendations for anaerobic joint infections. Cases involving both aerobes and anaerobes may require a two-drug regimen although the need to treat all of the cultivable potential pathogens is not established.

CHEST INFECTIONS

Anaerobic bacteria are well established and relatively common pulmonary pathogens. In our experience, these organisms are second only to *Streptococcus pneumoniae* in causing infections of the pulmonary parenchyma, and they are the most common isolates in empyema fluid. This experience stands in sharp contrast to that reported in the literature. It is our impression that the frequent failure to collect appropriate specimens, or to utilize optimal bacteriological methods, accounts for this discrepancy in bacteriological findings.

TYPES OF INFECTIONS

The major forms of anaerobic pleuropulmonary infections are aspiration pneumonitis, lung abscess, necrotizing pneumonia, empyema, and bronchiectasis. Anaerobes have been implicated in 50 to 90 per cent of patients in each of these disease categories when proper methods of study have been utilized (Table 7.2).

Table 7.2. Incidence of Anaerobic Pleuropulmonary Infection.

Literature Source	Condition studied	Cases	Cases yielding anaerobes
Beerens & Tahon-Castel 1965	Lung abscess	33	21 (64%)
Bartlett *et al.* (1974*b*)	Lung abscess	26	24 (92%)
Bartlett (*unpublished data*)	Necrotizing pneumonia	18	17 (94%)
Beerens & Tahon-Castel 1965	Empyema	46	23 (50%)
Bartlett *et al.* (1974*c*)	Empyema	83	63 (76%)
Swenson & Lerber (1974)	Aspiration pneumonia	40	22 (55%)
Bartlett *et al.* *(1974a)*	Aspiration pneumonia	70	61 (87%)

The most complete studies of anaerobic pleuropulmonary disease were performed in the pre-chemotherapy era (Smith, 1932; Kline & Berger, 1935; Pilot & Davis, 1924). At that time the natural course of the infection could be followed and surgical or autopsy specimens were often available to document the bacterial etiology.

In most cases aspiration was thought to be the underlying mechanism. This impression was based on the following evidence. Firstly, the bacteria implicated were the same as those that colonize the upper respiratory passages; secondly, many of the patients had anaerobic infections of the oropharynx, particularly dental infections; thirdly, typical lesions could be produced in experimental animals by the intratracheal inoculation of oral anaerobic bacteria (Smith, 1927); fourthly, most of the patients had underlying conditions which would predispose to aspiration, such as tonsillectomy or general anaesthesia, or other cause of altered consciousness; and finally, the most common pulmonary segments involved were those favoured by gravitational flow - the posterior segments of the upper lobes or the superior segments of the lower lobes in the recumbent position or the basal segments of the lower lobes in the upright position (Brock, 1952).

The sequence of clinical events in aspiration pneumonia was carefully traced by Smith (1932). Initially there was a pneumonitis, which was often associated with only minimal symptoms. Instead of proceeding to resolution, there was a tendency to abscess formation after 8-14 days. Cavitation was frequently accompanied by copious sputum production. X-rays taken at this stage showed a solitary, or dominant, relatively large pulmonary abscess ('primary lung abscess') or multiple small cavities within a pulmonary segment or lobe ('necrotizing pneumonia'). Pathological studies of lung abscesses obtained from surgically-resected specimens or autopsy material, showed a necrotic centre with multiple bacterial forms at the inflammatory front. With necrotizing pleumonia there were multiple adjacent small abscesses, sometimes accompanied by large tissue sloughs. Cultures obtained directly from these specimens yielded multiple anaerobic bacteria fusiforms, vibrios, streptococci, and spirochaetes.

In approximately 12 per cent of cases the infection extended to the pleural space either by contiguous spread or by bronchopleural fistula formation.

Although these patients often had indolent symptoms, the infection was associated with extensive morbidity and mortality. Symptoms often persisted for months or even years. Fatality rates ranged from 35-80 per cent, the highest rates occurring in patients with necrotizing pneumonia.

Our present concepts of anaerobic pleuropulmonary infection fit well with the clinical patterns described in these studies. Among 100 bacteriologically confirmed cases studied in our laboratory (Bartlett, Sutter & Finegold, 1974*d*) there were 44 cases of pneumonitis, 32 of lung abscess, and 24 of necrotizing pneumonia. There was an associated empyema in 41 cases. (This represents an obvious disproportion in the number of empyema cases, reflecting the ease of obtaining reliable specimens in the presence of this complication).

Among these 100 patients, 74 had associated conditions that predispose to aspiration as a result of compromised consciousness or dysphagia. The most common underlying conditions were alcoholism (35 patients), acute cerebrovascular accident (9), drug overdose (6), epilepsy (5), oesophageal diseases (5), intestinal obstruction with protracted vomiting (5) and general anaesthesia (4). Periodontitis or gingivitis was detected in 30 of the 62 patients who had dental consultations. X-ray evidence of involvement of the previously-noted dependent pulmonary segments provided further support for the impression that aspiration was a common antecedent event.

Although aspiration has been emphasized as the primary mechanism of these infections, other predisposing conditions were also noted. These include bronchogenic carcinoma (7 cases), bronchiectasis (6 cases), bland pulmonary embolism with infarction (9 cases), and empyema secondary to transdiaphragmatic spread of a subphrenic abscess (6 cases). Other investigators have also noted cases characterized by multiple sites of pulmonary infection following septic embolization from a distant focus of anaerobic suppuration.

The types of pathogenic processes and associated symptoms were largely determined by the stage of the disease at which patients presented. Cases detected early in the course (within 7-14 days of aspiration) usually showed radiological evidence of simple pneumonitis. Many patients with brief illnesses at the time of presentation had acute aspiration pneumonia, while lung abscess, necrotizing pneumonia, or empyema appeared to be relatively late complications. These patients usually had prolonged symptoms including cough, fever, dyspnoea, sputum production, and weight loss as well as anaemia. Approximately 50 per cent of patients had putrid sputum or empyema fluid. When present, this finding was usually noted relatively late in the course at the time abscesses were evacuated or empyemas formed.

These observations permit a formulation of certain clinical clues that have proved useful in preducting the presence of anaerobic bacteria in pleuropulmonary infections:

1. Most of the patients have observed or suspected aspiration as the underlying cause. Moreover, most cases of aspiration pneumonia are caused by anaerobic bacteria either alone or in combination with aerobic or facultative bacteria.
2. Putrid discharge is noted in approximately 50 per cent of cases and, when present, is considered diagnostic of anaerobic infection.
3. In contrast to pulmonary infections caused by aerobic and facultative bacteria, those involving anaerobes often begin insidiously and are characterized by a prolonged period of symptoms before presentation.
4. Pulmonary parenchymal cavitation or empyema formation are relatively common. In addition, most patients with lung abscesses, necrotizing pneumonia, or empyema have infections involving anaerobic bacteria.

BACTERIOLOGICAL METHODS

The above noted clues serve as guidelines to the clinical suspicion of anaerobic pleuropulmonary disease. A confirmed diagnosis, however, requires the recovery of anaerobic bacteria from a reliable specimen. Patients with empyema pose no problem in terms of a readily available specimen for meaningful anaerobic culture. When the infection is restricted to the pulmonary parenchyma, however, valid specimens are less easily obtained. It is the failure to study such patients that has been the greatest obstacle in detecting anaerobic pulmonary infections.

The upper respiratory tract harbours a rich flora of anaerobic bacteria, reflected by studies of saliva which yields concentrations of about 10^8 anaerobes/ml. The implication of this observation is that any specimen which is subject to salivary contamination is unsuitable for anaerobic culture. Thus, expectorated sputum is not valid and, in our experience, even quantitative cultures of these specimens fail to distinguish contaminants and pathogens (Bartlett & Finegold, 1971). In addition, aerobic cultures of expectorated sputum may be misleading by yielding potential pathogens that colonize the upper respiratory passages. It is the reliance on these specimens that accounts for the frequent reports incriminating *Staphylococcus aureus*, *Enterobacteriaceae*, *Pseudomonas aeruginosa* and other aerobic bacteria in aspiration pneumonia and cavitating pulmonary infections. In our view laboratories should not expend their energies in studying expectorated sputum when clinical clues suggest anaerobic infection.

Bronchoscopy aspirates collected in the usual manner (*i.e.* from the tip of the bronchoscope) and nasopharyngeal aspirates are also invalid for anaerobic culture. This is due to the contamination of the instrument or the catheter during passage through the upper airways. It is possible that the fibre-optic bronchoscope may prove useful for specimen collection, provided that special precautions are taken. Work from our laboratory clearly indicates that suction aspirates from the inner channel yield the oral flora. Specimen collection with a catheter passed through the inner channel and well beyond the instrument tip, may be reliable although this method has not been adequately studied. An additional precaution concerns the use of topical anaesthetics, which are usually employed with fibre-optic bronchoscopy. These agents are toxic to bacteria and their use may alter the cultivable flora (Schmidt & Rosenkranz, 1970).

At present the only specimens that can be used to study the bacteriology of anaerobic pleuropulmonary infections are those collected in a manner that completely bypasses the upper respiratory passages. These include pleural fluid, samples obtained by direct lung puncture, trans-tracheal aspirates, and pus collected at thoracotomy.

In our experience, the most useful specimen sources have been pleural fluid in patients with empyema, and trans-tracheal aspirates in cases where the infection is restricted to the pulmonary parenchyma. Recent studies have documented the validity of trans-tracheal aspirates for anaerobic culture (Bartlett, Rosenblatt & Finegold, 1973). This technique consists of passing a 16 gauge 'Intracath' through the cricothyroid membrane following subcutaneous anaesthesia. The catheter is passed to its full extent and the specimen is collected with a tight-fitting syringe or, preferably, with a Luken's trap and suction apparatus. Transtracheal aspiration is unpleasant for the patient because of the cough paroxysm that invariably results when the catheter is in the lower trachea. Contraindications to the procedure are severe hemoptysis, severe hypoxia, bleeding diathesis, and an uncooperative patient. Of 400 patients who have had transtracheal aspiration at our hospital, there have been two who developed severe subcutaneous emphysems, two who developed local abscesses at the needle puncture site, and one patient who had severe hemoptysis. All of these complications were self-limiting, and we regard the procedure as safe, with the proviso that the physician is familiar with the procedure and that patients with contraindications are not studied.

BACTERIOLOGICAL RESULTS

The infecting flora in anaerobic pleuropulmonary infections is usually complex (Table 7.3). Among the 100 patients studied in our laboratory, there were only 27 in whom a single anaerobic species was recovered in pure culture. The remaining 73 patients had an average of 3 isolates. Aerobic and facultative potential pathogens were present in combination with the anaerobes in 33 cases. The isolation of both types of bacteria inevitably raises questions regarding the true pathogens in these infections. In cases included in this study the anaerobes were the numerically predominant bacteria, and associated clinical features suggested that the anaerobic component was contributing to the pathogenic process. Admittedly, this is sometimes an arbitrary decision and reflects the ambiguities of precise definitions when one is dealing with mixed infections. It is of interest to note, however, that the clinical features, including the presence of putrid discharge, were equally prevalent in patients with only anaerobic isolates and those with anaerobes in combination with aerobes.

The predominant anaerobes in pulmonary infections are anaerobic and microaerophilic streptococci, *Fusobacterium nucleatum*, and *Bacteroides melaninogenicus* (Table 7.3). This reflects the flora of the upper respiratory passages and is consistent with the impression that the aspiration of oropharyngeal contents accounts for most cases.

The recovery of *Bacteroides fragilis* in 21 cases is of particular interest. Six of these patients had empyemas associated with subphrenic abscess - a microbiological relationship which is easily explained by the prevalence of *Bacteroides fragilis* in intra-abdominal infections. Aspiration was the suspected mechanism of infection in most of the remaining 15 patients. *B. fragilis* is not considered a normal inhabitant of the upper respiratory passages or gastric contents (Rosebury, 1962) and the

Table 7.3. Bacteriology of Anaerobic Pleuropulmonary Infections (100 cases).

	Cases
Only anaerobes recovered	66
Anaerobes in combination with aerobes	33
Average number of isolates per case - 2.5	
Anaerobic Gram-negative bacilli	
Fusobacterium nucleatum	34 (6)
Fusobacterium necrophorum	1
Bacteroides melaninogenicus	31 (1)
B. fragilis	21 (2)
B. oralis	16
Dialister pneumosintes	1
Unidentified	10
Anaerobic cocci	
Microaerophilic streptococcus	26 (12)
Peptostreptococcus	19 (3)
Peptococcus	15
Veillonella	7
Anaerobic Gram-positive bacilli	
Propionibacterium	9
Eubacterium	5
Bifidobacterium	3
Unidentified catalase-negative non-sporeforming	10 (1)
Clostridium perfringens	5 (2)
Clostridium species	2

Figures in parenthesis give number of cases with pure cultures.

source of this organism in such cases is obscure. The therapeutic implications of this finding are also of interest, since *B. fragilis* is the only anaerobe recovered with any frequency in these cases that is usually resistant to penicillin.

TREATMENT

The basic rules for treating anaerobic pleuropulmonary infections are the use of antimicrobials appropriate for the treatment of parenchymal infections, and adequate drainage of empyemas. Benzylpenicillin has long been regarded as the agent of choice for most cases. The vast majority of isolates are susceptible to penicillin, and extensive clinical use has corroborated these *in vitro* findings. The major exception is *B. fragilis* which was recovered from 21 per cent of cases as well as 15 per cent of patients with aspiration pneumonia (Bartlett, Gorbach & Finegold, 1974). It is conceivable that many of these patients would respond to penicillin since *B. fragilis* was generally recovered as part of a complex flora which also included other anaerobes. Thus, therapy directed against other components of the infection might prove satisfactory. Our data do not permit a conclusive answer to this question since, having isolated *B. fragilis*, we used appropriate agents.

There is sufficient evidence to conclude that clindamycin and chloramphenicol are equally efficacious, and they may be the agents of choice for infections involving *B. fragilis*. Unfortunately,there is insufficient clinical experience to judge the comparative efficacy of other antimicrobial agents in bacteriologically established cases.

Other therapeutic problems are the determination of the correct dose and duration of administration of antimicrobial agents. On the basis of published data and our own experience we would make the following recommendations.

Pneumonitis. Anaerobic pneumonitis can usually be treated in a manner analogous to pneumococcal pneumonitis. This usually consists of parenteral penicillin until clinical improvement is clearly noted, followed by several days of oral therapy. Patients usually become afebrile in 2-3 days and the chest X-ray clears after approximately 3 weeks.

Lung abscess. We generally treat patients with primary lung abscesses with a parenteral agent initially, although others have used large doses of oral penicillin right from the start with good results (Weiss, 1973). Our selection of parenteral therapy (particularly aqueous benzylpenicillin) is based on the concept that high serum concentrations may be required to obtain therapeutic levels in a relatively avascular necrotic lesion. When the patient is afebrile and clinically improved, we generally change to an oral agent. Treatment is continued until the chest X-ray is clear or shows a small stable residual lesion. This generally requires extended courses of treatment, much of which can be accomplished on an outpatient basis. The recommendation for prolonged therapy is based on our observation of five patients who relapsed when antimicrobials were discontinued prematurely. In our experience with 26 cases of primary lung abscess the mean duration of fever after the beginning of therapy was six days and the mean time for chest X-ray clearing was eight weeks. (Bartlett, Gorbach, Tally & Finegold, 1974*b*). An important finding in this study was that all patients had a satisfactory response to chemotherapy, provided that an appropriate agent was used according to the previous recommendations. No cases required surgery for the parenchymal infection. In our view surgery is specifically contraindicated since it may result in the intrabronchial spread of the infection.

Necrotizing pneumonia. Necrotizing pneumonia is a serious infection associated with a high mortality. Recommendations for antimicrobial therapy are the same as for lung abscess except that, with necrotizing pneumonia, there is a more compelling indication for the initial use of parenteral antimicrobials. Among 24 cases of this infection the mortality rate was 25 per cent, which emphasizes the need for early and aggressive chemotherapy.

Empyema. The most important therapeutic manoeuvre in patients with empyema is adequate drainage. Occasionally this can be accomplished by needle aspiration, but in most cases the fluid is purulent and loculated, requiring surgical drainage, usually with rib resection. In our experience with 41 cases of anaerobic empyema, five patients had closed surgical drainage and 32 required open drainage with rib resection. The remaining four patients had thoracocentesis only, and two of these died because of failure to achieve adequate drainage. The response to therapy in these cases was usually delayed and several patients required repeated surgical procedures. The mean duration of fever after the first drainage procedure was 18 days and the mean time required for drainage sites to heal was 13 weeks.

Mixed infections. Classical teaching is based on the concept that a single pathogen is responsible for the disease process. Therapy is simplified by selecting an agent that is active *in vitro* on the incriminated pathogen. With anaerobic infections this approach may be difficult to apply. In most of these cases the cultivable flora is complex and may include aerobes as well as anaerobes. The question that is raised concerns the need to direct therapy at all of the various potentially pathogenic components of the infection. This question has previously been posed regarding pulmonary infections involving *B. fragilis*. It is also frequently raised in cases involving aerobes in combination with anaerobes because the exact nature of these infections is ill-defined and precise answers are not available. Our approach has been to make therapeutic decisions on case-to-case basis. In serious and fulminating infections, we generally recommend therapy directed against all of the potential pathogens recovered. When patients are less severely ill the choice of chemotherapeutic agents may be influenced by the presence of putrid discharge, the relative numbers of bacteria in the Gram stain and in culture, the type of pathogenic process, and the specific bacteria isolated. If these suggest anaerobic infection, we have frequently used selective antimicrobial therapy - a single agent with activity against only the anaerobes. An analysis of 17 such cases showed no difference when compared with similar cases treated with a two-drug regimen. In both groups the duration of fever, the time required for radiological clearing, and the overall outcome, were similar. This must be viewed as a preliminary observation which requires more extensive investigation before firm conclusions can be drawn.

BACTERAEMIA

Recent studies from several institutions indicate that anaerobes account for 8-11 per cent of all bacteraemias. (Washington, 1971; Wilson *et al.*, 1972; Sonnenwirth, 1974; Chow & Guze 1974). These reviews include an aggregate total of 761 anaerobic isolates distributed as follows: 633 Gram-negative bacilli (84 per cent), 66 clostridia (9 per cent), 56 Gram-positive cocci (7 per cent) and 6 miscellaneous organisms (1 per cent). (Anaerobic blood culture isolates that were suspected contaminants have been deleted in this analysis).

The most common clinically significant anaerobe recovered in blood cultures is *Bacteroides fragilis*. This organism accounts for 70-90 per cent of the anaerobic Gram-negative bacillary blood culture isolates (Chow & Guze 1974; Wilson *et al.*, 1972; Nobles, 1973). This is consonant with the usual portals of entry in these cases and may also reflect current antibiotic usage. Wilson *et al.*, (1972) for example, noted that 56 per cent of their patients with anaerobic Gram-negative bacillaemia were receiving antimicrobial therapy at the time of positive blood cultures. The most commonly used agents were cephalosporins and ampicillin which are generally inactive against *B. fragilis*. At the Mayo Clinic *B. fragilis* is second only to *E. coli* in causing Gram-negative bacteraemia (Washington, 1971).

There appears to be a curious distribution of the five subspecies of *Bacteroides fragilis* in blood cultures. Approximately 80-90 per cent are subsp. *fragilis* while subsp. *thetaiotaomicron* accounts for 10-15 per cent. (Sonnenwirth, 1974). The implication of this observation in terms of declining pathogenic potential among the various subspecies has not been determined.

The portals of entry of most cases of anaerobic Gram-negative bacillaemia are the female genital tract or gastro-intestinal tract (see Chapter 8). Intraabdominal or pelvic surgery, gynaecological instrumentation, trauma, malignancies or other cause of a breach in the mucosal surface of the intestinal tract or female genital tract are common antecedant conditions. Decubitus ulcers, the oropharynx and the lung are less common portals of entry. In contrast with bacteraemia due to aerobic Gram-negative bacilli, the urinary tract is seldom the initiating focus of infection. This reflects the infrequency with which anaerobes are pathogens in cystitis and pyelonephritis (Segura *et al.*, 1972).

The clinical features of bacteraemia caused by anaerobic Gram-negative bacilli are often indistinguishable from those associated with their aerobic counterparts. Spiking fevers, systemic toxicity, and leucocytosis are usually present. Gram-negative anaerobes possess endotoxin (Sonnenwirth *et al.*, 1972) and the septic shock syndrome has been noted in 20-35 per cent of bacteraemic patients (Table 7.4). A similar incidence of septic shock is reported in most studies of bacteraemia involving aerobic Gram-negative bacilli. Distinctive features of anaerobic bacteraemia (Table 7.4) are: the previously mentioned portals of entry, an excessive incidence of jaundice, and a tendency to septic thrombophlebitis with distant septic embolization which may prove refractory to heparin therapy because of heparinase production by some anaerobic strains (Gesner & Jenkin, 1961). In some cases surgical intervention is often required to prevent continuing embolization and a lethal outcome.

Table 7.4. Clinical Features of Anaerobic Gram-negative Bacillaemia

	Cases	Septic Shock	Jaundice	Thrombophlebitis	Septic embolization	Mortality
Chow & Guze (1974)	112	35(31%)	N.S.*	6(5%)	30(27%)	35(31%)
Gelb & Seligman (1970)	35	12(34%)	10(29%)	N.S.	N.S.	13(37%)
Wilson *et al.* (1972)	52	19(37%)	7(13%)	23(44%)	N.S.	16(31%)
Nobles (1973)	43	7(16%)	5(12%)	4(9%)	10(23%)	15(35%)
Marcoux *et al.* (1970)	123	27(22%)	12(10%)	6(5%)	28(23%)	35(29%)

* N.S. = Not specified

Mortality rates among all cases are consistently 25-35 per cent, although there are important differences within individual patient populations. In a review of 154 reported anaerobic bacteraemias in obstetric patients there was only a single fatality (Chow & Guze, 1974). In contrast, the mortality rate among patients with an ultimately fatal underlying disease was 58 per cent. Chemotherapy is an additional important prognostic factor. In two studies the mortality rate was reduced from 60 per cent to 12-16 per cent when the antibiotic administered was appropriate for the blood culture isolate (Nobles, 1973; Chow & Guze, 1974). It is worth recalling that 81 per cent of anaerobic Gram-negative bacteraemias were fatal in the pre-chemotherapeutic era (Gunn, 1956). These data emphasize the importance of early recognition and appropriate antimicrobial treatment.

Anaerobes other than Gram-negative bacilli account for 10-30 per cent of all anaerobic bacteraemias and 1-3 per cent of all septicaemias. *Peptostreptococcus*, *Peptococcus* and clostridia are the major isolates (Washington, 1971; Wilson *et al.*, 1972; Sonnenwirth, 1974). In these cases the initiating focus of infection is generally the female genital tract, oropharynx, liver abscess, gastrointestinal tract or soft tissue.

Clostridial septicaemias are of particular interest because physicians often associate them with a devastating course characterized by necrotic tissue, haemolytic anaemia and renal failure. Although this is a well established syndrome it accounts for a relatively small portion of all cases of clostridial bacteraemia. This was noted in the pre-chemotherapeutic era. Twenty-eight obstetric patients with *Clostridium perfringens* bacteraemia had minimal toxicity and all survived despite the lack of chemotherapy (Ramsey, 1949). A similar observation was also made more recently by Rathbun (1968). Twenty medical and surgical patients with bacteraemia caused by histotoxic clostridia had no evidence of hemolysis or gas gangrene. The lesson from these studies is that the therapy and prognosis in cases of clostridial bacteraemia should be evaluated together with the associated clinical findings.

The therapy of anaerobic bacteraemia is based on the usual principles of surgical drainage of pus and appropriate chemotherapy. There is good *in vitro* and *in vivo* data to show that chloramphenicol and clindamycin are excellent agents for *B. fragilis* bacteraemia (Martin *et al.*, 1972; Chow & Guze, 1974). Benzylpenicillin is regarded as the agent of choice in most patients with bacteraemia involving anaerobes other than *B. fragilis*, but in such cases clindamycin and chloramphenicol also serve as alternative agents.

ENDOCARDITIS

In two recent reviews of endocarditis, which include an aggregate total of 1498 cases, anaerobes were recovered in 55 (3 per cent) (Felner & Dowell, 1970; Weinstein & Rubin, 1973). Anaerobic and microaerophilic streptococci accounted for 53 of these. It is likely that the incidence of oxygen-sensitive forms would be greater if improved culture techniques were used. This hypothesis is based on the fact that approximately 15 per cent of patients with endocarditis have persistently negative blood cultures, and the relatively frequent occurrence of anaerobic bacteraemia associated with oral surgery (deVries, Francis & Lang, 1972).

Additionally, the incidence of microaerophilic streptococci in endocarditis may commonly be underestimated. This results from the tendency of many laboratories to group these organisms with facultative streptococci. Several investigators have implicated microaerophilic streptococci in as many as 8-16 per cent of cases (Table 7.5) (Dormer, 1958; Blount, 1965; Lerner & Weinstein, 1966; Tompsett, 1967; Shinebourne *et al.*, 1969).

TABLE 7.5. Incidence of Microaerophilic Streptococci in Endocarditis.

Literature source	Cases	Cases involving microaerophilic streptococci
Dormer (1958)	82	8 (10%)
Blount (1965)	89	7 (8%)
Lerner & Weinstein (1966)	100	13 (13%)
Tompsett (1967)	77	7 (9%)
Shinebourne *et al.* (1969)	63	10 (16%)

The clinical features of endocarditis involving anaerobic and microaerophilic streptococci are generally analogous to those caused by *Streptococcus viridans*. The course is generally subacute, most of the patients have predisposing valvular disease, and an oral portal of entry is usually suspected. Intravenous penicillin for four weeks generally constitutes adequate therapy, although this should be supported by susceptibility tests demonstrating bactericidal activity at achievable levels.

Other anaerobic bacteria that may rarely cause endocarditis include *Bacteroides*, fusobacteria, clostridia, propionibacteria, *Veillonella*, *Actinomyces* and eubacteria. In these cases the same therapeutic principles apply.

Nastro and Finegold (1973) reviewed 37 published reports of endocarditis due to anaerobic Gram-negative bacilli and contributed an additional case involving *B. fragilis*. The most commonly identified isolates in this review were *B. fragilis* (13 cases), *Fusobacterium necrophorum* (7) and *Fusobacterium nucleatum* (3). Most of the cases were associated with a subacute course although 2 patients with *Fusobacterium necrophorum* endocarditis had a fulminant disease terminating fatally within 10 days. Underlying cardiac disease was present in 33 (64 per cent) and there was a high incidence of major embolic complications (67 per cent). The mortality rate among cases reported in the chemotherapeutic era was 38 per cent.

Patients with *Bacteroides fragilis* endocarditis often present a special therapeutic problem in that most antimicrobials are either inactive against this organism or they are only bacteriostatic *in vitro*. Metronidazole is bactericidal (Nastro & Finegold, 1973) although there is minimal clinical experience with this agent and no parenteral form is currently available.

REFERENCES

Bartlett, J.G. & Finegold, S.M. (1971). Improved technique to process coughed sputum for culture. *Clinical Research*, 19, 183.

Bartlett, J.G., Rosenblatt, J.E. & Finegold, S.M. (1973). Percutaneous transtracheal aspiration in the diagnosis of anaerobic pulmonary infections. *Annals of Internal Medicine*, 79, 535.

Bartlett, J.G., Gorbach, S.L. & Finegold, S.M. (1974*a*). Bacteriology of aspiration pneumonia. *American Journal of Medicine*, 56, 202.

Bartlett, J.G., Gorbach, S.L., Tally, F.P. & Finegold, S.M. (1974*b*). Bacteriology and treatment of primary lung abscess. *American Review of Respiratory Diseases* (in press).
Bartlett, J.G., Gorbach, S.L., Thadepalli, H. & Finegold, S.M. (1974*c*). Bacteriology of empyema. *Lancet*, 1, 338.
Bartlett, J.G., Sutter, V.L. & Finegold, S.M. (1974*d*). Anaerobic pleuropulmonary disease, clinical observations and bacteriology in 100 cases. *Anaerobic Bacteria: role in disease*. Ed. Balows, A., De Haan, R.M., Dowell, V.R. Jr. & Guze, L.B. Charles C. Thomas, Springfield, Illinois. p.327.
Beerens, H. & Tahon-Castel, M. (1965). Infection Humaines à Bactéries, Anaérobies Non-toxigènes. Presses Académiques Européennes, Bruxelles, p.92.
Brock, R.C. (1952). Lung Abscess. Charles C. Thomas, Springfield, Ill.
Blount, J.G. (1965). Bacterial endocarditis. *American Journal of Medicine*, 38, 909.
Chernik, N.L., Armstrong, D. & Posner, J.B. (1973). Central nervous system infections in patients with cancer. *Medicine*, 52, 563.
Chow, A.W. & Guze, L.B. (1974). Bacteroidaceae bacteremia: Clinical experience with 112 patients. *Medicine* (In press).
Coffey, J.D., Jr. (1966). Otitis media in the practice of paediatrics. *Pediatrics*, 38,25.
Cohen, I. (1934). Cerebral complications of putrid pleuropulmonary suppuration. *Archives of Neurology and Psychiatry*, 32, 174.
DeVries, J., Francis, L .E. & Lang, D. (1972). Control of post-extraction bacteremias in the penicillin-hypersensitive patient. *Journal of Canadian Dental Association*, 38, 63.
Dormer, A.E. (1958). Bacterial endocarditis: Survey of patients treated between 1945 and 1956. *British Medical Journal*, 1, 63.
Feingold, M., Klein, J.D., Halsam, G.E., Tilles, J.G. Finland, M. & Gellis, S.S. (1966). Acute otitis media in children. *American Journal of Diseases of Children*, 111, 361.
Felner, J.M. & Dowell, V.R., Jr. (1970). Anaerobic bacterial endocarditis. *New England Journal of Medicine*, 283, 1188.
Felner, J.M. & Dowell, V.R., Jr. (1971). 'Bacteroides' bacteremia. *American Journal of Medicine*, 50, 787.
Finegold, S.M., Posnick, D.J., Miller, L.G. & Miller, A.B. (1966). Sensitivity of major bacteroides groups to antibacterial agents. *Bacteriological Proceedings*, p.64.
Finegold, S.M., Suglhara, P.T. & Miller, A.B. (1968). Susceptibility of anaerobic cocci to antibacterial agents. *Bacteriological Proceedings, 1967*, p.96.
Frederick, J. & Braude, A.I. (1974). Anaerobic infection of the paranasal sinuses. *New England Journal of Medicine*, 290, 135.

Garrod, L.P. & O'Grady, F. (1972). Antibiotic and Chemotherapy. Churchill Livingstone: Edinburgh and London, p.138.
Geddes, A.M., Sleet, R.A. & Murdoch, J.M. (1964). Lincomycin hydrochloride: Clinical and laboratory studies. *British Medical Journal*, 2, 670.
Gelb, A.F. & Seligman, S.J. (1970). Bacteroidaceae bacteremia: Effect of age and focus of infection upon clinical course. *Journal of the American Medical Association*, 212, 1038.
Gesner, B.M. & Jenkin, C.R. (1961). Production of heparinase by bacteroides. *Journal of Bacteriology*, 81, 595.
Gregory, D.H., Messner, R. & Zinneman, H.H. (1967). Metastatic brain abscesses: A retrospective appraisal of 29 patients. *Archives of Internal Medicine*, 119, 25.
Gunn, A.A. (1956). Bacteroides septicemia. *Journal of the Royal College of Surgeons, Edinburgh*, 2, 41.
Heineman, H.S. & Braude, A.I. (1963). Anaerobic infection of the brain. *American Journal of Medicine*, 35, 682.
Howie, V.M., Ploussard, J.H. & Lester, R.L., Jr. (1970). Otitis media: A clinical and bacteriological correlation. *Pediatrics*, 45, 29.
Kelly, P.J., Martin, W.J. & Coventry, M.H. (1970). Bacterial (suppurative) arthritis in the adult. *Journal of Bone and Joint Surgery*, 52A, 1595.
Kelly, P.J., Wilkowske, C.J. & Washington, J.A.,II, (1973). Comparison of Gram-negative bacillary and staphylococcal osteomyelitis of the femur and tibia. *Clinical Orthopedics and Related Research*, 96, 70.
Kline, B.S. & Berger, S.S. (1935). Pulmonary abscess and pulmonary gangrene. *Archives of Internal Medicine*, 56, 753.
Lemierre, A. (1936). On certain septicaemias due to anaerobic organisms. *Lancet*, 340, 701.
Lerner, P.I. & Weinstein, L. (1966). Infective endocarditis in the antibiotic era. *New England Journal of Medicine*, 274, 199.
MacDonald, J.B., Socransky, S.S. & Gibbons, R.J. (1963). Aspects of the pathogenesis of mixed anaerobic infections of mucous membranes. *Journal of Dental Research*, 42, 529.
Mangi, R.J., Kundargi, R.S., Quintiliani, R. & Andriole, V.T. (1973). Development of meningitis during cephalothin therapy. *Annals of Internal Medicine*, 78, 347.
Marcoux, J.A., Zabransky, R.A., Washington, J.A.,II, Wellman, W.E. & Martin, W.J. (1970). Bacteroides bacteremia: A review of 123 cases. *Minnisota Medicine*, 53, 1169.
McFarlan, A.M. (1943). The bacteriology of brain abscess. *British Medical Journal*, 2, 643.
Martin, W.J., Gardner, M. & Washington, J.A.,II, (1972). In vitro antimicrobial susceptibility of anaerobic bacteria isolated from clinical specimens. *Antimicrobial Agents and Chemotherapy*, 1, 148.

Nastro, L.J. & Finegold, S.M. (1973). Endocarditis due to anaerobic Gram-negative bacilli. *American Journal of Medicine*, 54, 482.
Newton, E.J. (1956). Haematogenous brain abscess in cyanotic congenital heart disease. *Quarterly Journal of Medicine*, 25, 201.
Nobles, E.R. (1973). Bacteroides infections. *Annals of Surgery*, 177, 601.
Pilot, I. & Davis, D.J. (1924). Studies in fusiform bacilli and spirochetes. *Archives of Internal Medicine*, 34, 313.
Ramsey, A.M. (1949). The significance of *Clostridium welchii* in the cervical swab and blood stream in postpartum and postabortion sepsis. *Journal of Obstetrics and Gynaecology of the British Empire*, 56, 247.
Rathbun, H.K. (1968). Clostridial bacteremia without hemolysis. *Archives of Internal Medicine*, 122, 496.
Rosebury, T. (1962). Microorganisms Indigenous to Man. McGraw-Hill Book Co., Inc., New York.
Schmidt, R.M. & Rosenkranz, H.S. (1970). Antimicrobial activity of local anesthetics: lidocaine and procaine. *Journal of Infectious Disease*, 121, 597.
Segura, J.W., Kelalis, P.P., Martin, W.J. & Smith, L.H. (1972). Anaerobic bacteria in the urinary tract. *Mayo Clinic Proceedings*, 47, 30.
Shinebourne, E.A., Cripps, C.M., Hayward, G.W. & Shooter, R.A. (1969). Bacterial endocarditis 1956-1965: Analysis of clinical features and treatment in relation to prognosis and mortality. *British Heart Journal*, 31, 536.
Smith, D.T. (1927). Experimental aspiratory abscess. *Archives of Surgery*, 14, 231.
Smith, D.T. (1932). Oral Spirochetes and Related Organisms in Fusospirochetal Disease. The Williams & Wilkins Co., Baltimore.
Sonnenwirth, A.C., Yin, E.T., Sarmiento, E.M. & Wessler, S. (1972). Bacteroidaceae endotoxin detection by Limulus assay. *American Journal of Clinical Nutrition*, 25, 1452.
Sonnenwirth, A.C. (1974). Incidence of intestinal anaerobes in blood cultures. In *Anaerobic Bacteria: role in disease*. Ed. Balows, A., De Haan, R.M., Dowell, V.R. Jr. & Guze, L.B. Charles C. Thomas, Springfield, Illinois. p.157.
Swartz, M.N. & Karchmar, A.W. (1974). Infections of the central nervous system. In *Anaerobic Bacteria: role in disease*. Ed. Balows, A., De Haan, R.M., Dowell, V.R. Jr. & Guze, L.B. Charles C. Thomas, Springfield, Illinois. p.309.
Swenson, R.M. & Lorber, B. (1974). Bacteriology of aspiration pneumonia: Community versus hospital acquired. *Clinical Research*, 22, 38A.

Tompsett, R. (1967). Bacterial endocarditis: Changes in the clinical spectrum. *Archives of Internal Medicine*, 119, 329.

Waldvogel, F.A., Medoff, G. & Swartz, M.N. (1970). Osteomyelitis: A review of clinical features, therapeutic considerations and unusual aspects. *New England Journal of Medicine*, 282, 198.

Washington, J.A.,II. (1971). Comparison of two commercially available media for detection of bacteremia. *Applied Microbiology*, 22, 604.

Weinstein, L. & Rubin, R.H. (1973). Infective endocarditis 1973. *Progress in Cardiocascular Diseases*, 16, 239.

Weiss, W. (1973). Cavity behavior in acute, primary, nonspecific lung abscess. *American Review of Respiratory Diseases*, 108, 1273.

Williams, A.C. & Gurainick, W.C. (1943). The diagnosis and treatment of Ludwig's angina A report of twenty cases. *New England Journal of Medicine*, 228, 443.

Wilson, W.R., Martin, W.J., Wilkowske, C.J. & Washington, J.A.,II. (1972). Anaerobic bacteremia. *Mayo Clinic Proceedings*, 47, 639.

Ziment, I., Davis, A. & Finegold, S.M. (1969)..Joint infection by anaerobic bacteria: A case report and review of the literature. *Arthritis and Rheumatism*, 12, 627.

Ziment, I., Miller, L.G. & Finegold, S.M. (1968). Nonsporulating anaerobic bacteria in osteomyelitis. *Antimicrobial Agents and Chemotherapy - 1967*, p.77.

DISCUSSION

DR PHILLIPS (LONDON): May I ask Dr Bartlett about the response of lung abscesses in particular to chemotherapy alone? Did you have to intervene surgically with these?

DR BARTLETT: I'm glad you asked the question because I meant to include it in the discussion. Our experience with lung abscess now is with a total of 45 patients and I think the series is somewhat remarkable in that not a single patient required surgical drainage - in fact, we feel surgical drainage of lung abscesses is contraindicated. This incurs the risk of spillage of the abscess contents into other parts of the lung. All of these patients responded to chemotherapy once we identified the bacterial flora and specifically directed our therapy at the organisms that were recovered. The mean duration of fever after the inception of therapy was 6 days, cavity closure often required 4-6 weeks but they all eventually closed and the Roentgenograms also universally cleared. In the entire experience there were only a few patients that had response that was not as I described and those were patients who had concurrent illnesses which explained their somewhat different course. Our impression is that primary lung abscess, and not those lung abscesses which are complications of other conditions, will universally respond to appropriate antimicrobial agents. We continue antibiotics until the X-ray is clear or shows a small stable residue. The reason for prolonging therapy is that we found that when we stopped somewhat short of that, many of the patients had either a relapse, with recurrence of fever or the reappearance of cavity on chest X-ray. We now give them rather intensive therapy in the hospital until they defervesce and show clinical improvement and then continue for extended periods of time - often 8 or 10 weeks - with oral agents on an outpatient basis.

DR GHONEIM (EXETER): In cases of endocarditis, is there any possibility of this happening in normal endocardium or are these damaged valves? Secondly, about the chemotherapy of cases of endocarditis and bacteraemia caused by *Bacteroides fragilis* is there any possibility for metronidazole to replace clindamycin?

DR BARTLETT: Right now there is not very much good clinical experience with metronidazole for anaerobic infections and I really am not prepared to say whether it is a good agent for anaerobic bacteraemia. There are some preliminary studies from Dr Finegold and his co-workers which suggest that it is a promising agent. One of the big drawbacks of metronidazole at this point is that it is not available parenterally and some people would be reluctant to use an oral agent for a severely ill patient. I think its greatest potential use is in *Bacteroides fragilis* endocarditis - although those cases are quite rare. We do have alternative parenteral agents for *Bacteroides fragilis* bacteraemia, but metronidazole is the only agent that is regularly bactericidal against this organism. Endocarditis would be the special circumstance in which you might specifically want metronidazole. Most of the cases of anaerobic endocarditis I mentioned were due to microaerophilic and anaerobic *Streptococcus*. They tend to occur in valves that have been damaged but they can certainly occur in valves that have not been damaged. If you exclude those cases and consider endocarditis caused by anaerobic Gram-negative rods, only about half of the patients have had previous valve damage (Nastro & Finegold, 1973).

PROF. GARROD: Could I briefly ask Dr Bartlett two questions? One is that he seems to assume that some of these septic lung lesions are the result of aspiration but surely some of the lung abscesses are the result of septic embolisation from a dental focus? I have always taken it that that is the usual origin of pulmonary actinomycosis, that is to say blood borne and not aspiration. Secondly, I was puzzled at the inclusion of sinusitis in his presentation - I would have thought that anaerobic infection of sinusitis could only have arisen from infection from a tooth in the upper jaw into the maxillary sinus. Were all these cases maxillary or how do you account for the fact? The usual causes of sinusitis are the pyogenic cocci, and *Haemophilus* of course.

DR BARTLETT: I think the sinusitis that you are referring to is probably acute sinusitis which is usually caused by other organisms. In fact, we have cultured acute sinusitis fluid and been able to recover only a few anaerobes and those that we did recover, we felt were often contaminants. There are some other workers who have done the same with similar results - that anaerobes tend to be seldom encountered in acute sinusitis but may be involved in chronic sinusitis. The anaerobes in chronic sinusitis presumably originate from the nose and gain access to the sinuses as a result of chronic inflammation or perhaps they are there from the beginning. With reference to your first question, we assume that most lung abscesses occur as a result of oral-pharyngeal contamination of the lower respiratory tract by the mechanism of aspiration rather than by a blood borne mechanism. We would suspect that if it was a blood-borne infection there might be multiple foci rather than just a single focus. Although there is good evidence that people tend to aspirate periodically those people who develop lung abscesses are usually subject to repeated bouts of aspiration or aspiration of large volumes, such as alcoholics or patients who have general anaesthesia. I would assume that the relationship of dental infection to lung abscess is one in which the dental infection accounts for rather rich flora of anaerobes in the inoculum that drains down the tracheal and bronchial tree. I don't think that the blood-borne infection really accounts for very many of the cases but I agree that we really don't know the true mechanism in some cases.

DR HARRISON (BRISTOL): As the lung abscesses which Dr. Bartlett referred to did not have surgical treatment, could he please tell us how he obtained the specimens for bacteriological examination, were they from sputum, and if so, was it processed in any way?

DR BARTLETT: No, we collected all of the specimens from lung abscess by trans-tracheal aspiration by the method that I mentioned. All of the bacteriological samples in all the patients were obtained from either blood, empyema fluid, or trans-tracheal aspiration, thoracotomy specimens, or direct lung puncture. We have never used expectorated sputum to diagnose anaerobic pulmonary infection. We have not been able to develop techniques to make expectorated sputum reliable for anaerobic culture. In fact we have found these specimens very confusing because they frequently yield misleading aerobic bacteria.

REFERENCE

Nastro, L.J. & Finegold, S.M. (1973). Endocarditis due to anaerobic Gram-negative bacilli. *American Journal of Medicine*, 54, 482.

8. Intra-abdominal, Genito-urinary, Skin and Soft Tissue Infections due to Non-sporeforming Anaerobic Bacteria

S. M. FINEGOLD

INTRA-ABDOMINAL AND PERINEAL INFECTIONS

Considering that anaerobic bacteria account for over 99 per cent of the normal colonic flora (Finegold, 1969), it is surprising that they have not been reported more frequently from intra-abdominal infections. Clearly this relates to problems of specimen collection, transport, and conditions of culture (see Chapters 2 and 3). A recent study of 46 patients with various types of intra-abdominal sepsis by Gorbach, Thadepalli & Norsen (1973) confirms that anaerobes *do* play a major role in such infections. In 43 cases, specimens of pus from the abdomen were available: 33 contained mixtures of aerobes and anaerobes, seven contained anaerobes alone, and three had only aerobic or facultative organisms. The other three patients had positive blood cultures only. A total of 104 anaerobes and 88 aerobic or facultative organisms was recovered, an average of five isolates (three anaerobes and two aerobes) per specimen, and a range of 1 to 13 organisms per sample. *Bacteroides fragilis* was the commonest anaerobe (28 isolates) and there were 8 isolates of other *Bacteroides* species, 6 *Fusobacterium*, 11 *Peptostreptococcus*, 3 *Peptococcus*, 1 *Veillonella*, 31 *Clostridium*, 11 *Eubacterium*, 2 *Propionibacterium*, 2 *Lactobacillus*, and 1 *Bifidobacterium*. The major facultative organism recovered was *Escherichia* (present 28 times). Positive blood cultures were found in 13 of the 46 patients; 11 of these had only anaerobes and 2 had both aerobes and anaerobes. In the patients with bacteraemia as well, *Bacteroides fragilis* was again the predominant organism (8 out of the 13).

Also of interest and an indication of the importance of anaerobes are the preliminary studies of Thadepalli and Gorbach, (1972) in which 12 patients with mixed aerobic and anaerobic abdominal and pelvic infections were treated with clindamycin alone. This antibiotic, despite lack of activity against such facultative forms as *E. coli* and *Strep. faecalis* gave good clinical results.

Other work demonstrates the importance of anaerobes but presumably because of limitations of techniques, the profusion of organisms reported by Gorbach *et al.* (1973) has not usually been found. Weiss (1943) reported the recovery of *Bacteroides melaninogenicus* from ten intra-abdominal infections, while Lodenkämper & Stienen (1955) reported 12 pure *Bacteroides* infections and later (Lodenkämper & Stienen, 1956) noted that among 722 anaerobes isolated from various infectious processes, 137 were from the abdominal area. Stokes (1958) found that approximately one-third of 528 specimens of abdominal pus that yielded a positive culture yielded anaerobes, 30 of them in pure culture. Bornstein *et al.* (1964) reported an infected mesenteric cyst containing five litres of fluid, which yielded a pure culture of *Bacteroides*. Saksens *et al.* (1968) reported on the role of *Bacteroidaceae* in 112 surgical infections, sixty of which were related to the gastrointestinal tract. Finally Armfield *et al.* (1971) found that *Clostridium innocuum* and *Eubacterium filamentosum* were usually isolated from infections involving the gastrointestinal tract.

Analysis of cases of anaerobic bacteraemia also underlines their importance in intra-abdominal sepsis. Gunn (1956) reviewed 148 cases of *Bacteroides* bacteraemia and noted that 26 were related to abdominal lesions, while in a series of 123 cases of *Bacteroides* bacteraemia, Marcoux *et al.* (1970) found that the gastrointestinal tract was the most common portal of entry, accounting for 61.8 per cent of the bacteraemias. Bodner, Koenig & Goodman (1970) reported that 24 of their 39 cases of *Bacteroides* bacteraemia originated in the gastrointestinal tract. Kuklinca & Gavan (1971) reported on 237 postmortem blood cultures, 126 of which were sterile. There were 16 anaerobic isolates, all but one from patients with gastrointestinal lesions.

The association of anaerobic abdominal infection and carcinoma of the bowel is a common and important one. This type of infection may be the first manifestation of the malignancy. Finegold, Marsh & Bartlett (1971) reported 13 cases of anaerobic infection in association with carcinoma of the colon and mentioned an additional 58 cases from a review of the literature. Additional cases have been reported by a number of other workers. In the paper by Marcoux *et al.* (1970) 21 cases of carcinoma of the colon were reported and in the report of Saksena *et al.* (1968) there were 18 such cases. Bornstein *et al.* (1964) report 6 cases of *Bacteroides* infection associated with carcinoma of the bowel, as well as some intra-abdominal abscesses due to anaerobic streptococci that were related to bowel carcinoma. In the series of bacteraemia patients reported by Wilson *et al.* (1972) 23 per cent of 52 patients with septicaemia due to anaerobic Gram-negative bacilli had carcinoma of the colon.

The important role of anaerobes in infections related to penetrating abdominal trauma is emphasized by the studies of Thadepalli *et al.* (1972, 1973). Other bowel lesions (vascular, obstructive, inflammatory, *etc.*) may also be the source of anaerobic infection. Fifteen per cent of the 52 patients with anaerobic Gram-negative rod bacteraemia in the series of Wilson *et al.* (1972) had small bowel perforation as the underlying source of the bacteraemia.

One further type of anaerobic infection that may be associated with the gut is actinomycosis. The most common type of abdominal actinomycosis is that involving the ileocecal region. This disease results in firm, woody masses within the abdomen and abscess formation, and draining sinuses and fistulae are common. Because of the firm masses that are produced, these lesions are readily confused with carcinoma. Harvey, Cantrell & Fisher (1957) described 10 cases of actinomycosis originating in the gastrointestinal tract.

POST-OPERATIVE AND OTHER WOUND INFECTIONS

Anaerobes play a major role in post-operative infections related to the abdomen. Spaulding *et al.* (1971) cultured 55 post-operative wounds, 45 of which yielded one organism or another. Among these, 37 were anaerobic organisms (only one in pure culture) and a total of 82 strains of anaerobes, representing 38 different genera or species, was recovered, Most numerous was *Bacteroides fragilis* (recovered 18 times), followed by *Bacteroides melaninogenicus* (10), *Peptococcus prevotii* (7) *Fusobacterium nucleatum* (6),and *F. ridiculosum* (5). Among a group of 301 cases of post-operative wound infection seen in a period of somewhat over two years, Hoffman & Gierhake (1969) found anaerobes involved in 33 per cent. In this study, cultures were obtained by needle puncture of wounds which were still closed. Members of the *Bacteroides* group and anaerobic streptococci were found with approximately equal frequency and greatly outnumbered all other types of anaerobes recovered. In 69 per cent of the infections in which anaerobes were involved, aerobic or facultative bacteria were found in addition, though usually in much smaller numbers. In another study, by Azar & Drapanas (1968) *Bacteroides* species accounted for 29 per cent of wound infections following elective surgery on the colon. It is interesting to note in this paper that cultures of the bowel at the time of surgery in 29 patients who were prepared with castor oil, a low residue diet, and 6 to 10 doses of oral neomycin and sulfathalidine, yielded *Bacteroides* species in 10 cases and *Clostridium* species in two. There are several other studies that also demonstrate the importance of anaerobes in infections of post-operative abdominal wounds. (Sandusky, 1942; Gillespie & Guy, 1956; Saksena *et al.*, 1968; Beazley *et al.*, 1972) Bernard & Cole (1963) advocated delayed primary wound closure following potentially contaminated operations in order to minimize the risk of anaerobic infection. They noted that 5 of 33 patients who had primary wound closure developed serious would infections, 3 of which involved anaerobes, whereas none of 40 patients having delayed primary wound closure developed infection.

ABDOMINAL WALL INFECTIONS

Abdominal wall infections may follow a variety of infectious, traumatic or other pathological processes within the abdomen, or involving the female genitalia. Sandusky *et al.* (1942) described 5 cases of abdominal wall abscess due to anaerobic streptococci mixed with other organisms, including *Actinomyces*, *Bacteroides*, and other anaerobic Gram-negative bacilli, while Lodenkämper & Stienen (1955) reported eight abdominal wall abscesses yielding pure cultures of *Bacteroides* and 2 additional abscesses with mixed cultures involving *Bacteroides* and other organisms. Heinrich & Pulverer (1960) reported the isolation of *Bacteroides melaninogenicus* from 11 cases of abdominal wall abscess, and Furuta & Tsuchiya (1973) reported 7 abscesses of the abdominal wall, all of which yielded *Bacteroides fragilis* (5 times in pure culture). Six of the patients in this last series had had an appendicectomy prior to their infection. Harvey, Cantrell & Fisher (1957) reported 15 cases of actinomycosis with secondary spread to the abdominal wall.

SUBPHRENIC AND OTHER INTRAPERITONEAL ABSCESSES

In the excellent paper by Altemeier *et al.* (1973), which included 501 patients with 540 intra-abdominal abscesses, most of which were intra-peritoneal, 19 per cent originated with appendicitis, 8 per cent with lesions of the biliary tract, and 7 per cent with diverticulitis. In this study, at least half of the bacteria found in intra-abdominal abscesses were anaerobic. The authors noted that when careful technique was used the frequency of anaerobic pathogens was even greater (60-70 per cent). Relating the bacteriological findings to the primary disease, they found that *Bacteroides* was recovered from 17 per cent of cases of intra-abdominal abscesses secondary to appendicitis, 44 per cent of those related to the female genitourinary tract, 27 per cent to the male genitourinary tract, 20 per cent to diverticulitis and 24 per cent to trauma. Anaerobic streptococci were found in 25 per cent of intra-abdominal abscesses where the primary disease was appendicitis, 38 per cent related to female genitourinary tract disease, 23 per cent to male genitourinary tract disease, 11 per cent to diverticulitis and 6 per cent to trauma. Magilligan (1968) found an incidence of 5 per cent of *Clostridium perfringens (welchii)* and 3 per cent of anaerobic streptococci among 82 cases of subphrenic abscess; however, 14 per cent of abscesses were sterile on culture and 10 per cent yielded *Staphylococcus epidermidis*, an obvious contaminant. Wilson *et al.* (1972) found that 8 per cent of 52 patients with anaerobic Gram-negative bacillus bacteremia had intra-abdominal abscesses. Bornstein *et al.* (1964) reported 8 intra-abdominal abscesses due to anaerobic streptococci, all of which were related to either bowel or gallbladder surgery. Others have also found anaerobes in intraperitoneal abscesses (Sandusky *et al.* 1942; Shoemaker, 1960). Harvey *et al.* (1957) reported four cases of subphrenic abscess due to *Actinomyces*.

RETROPERITONEAL ABSCESSES

There are a number of individual case reports of retroperitoneal abscess with a variety of underlying lesions including appendicitis, pulmonary or pleural infection and septic abortion. The organisms involved include *Bacteroides*, *Actinomyces israelii*, *Fusobacterium (Sphaerophorus) necrophorum*, fusiform bacilli and anaerobic streptococci.

PERITONITIS

There are a number of causes of peritonitis, many of which predispose to infection with anaerobes. Included in the latter category are diseases and injuries of the gastrointestinal tract, lesions of the female genital system, lesions of the biliary tree, and abdominal surgery, particularly surgery on the gastrointestinal tract. Multiple strains of bacteria are commonly isolated from cases of peritonitis and the more organisms present in the peritoneal exudate, the worse the pyognosis. Altemeier (1938), in a classic paper, describes the bacterial flora of peritonitis secondary to acute appendicitis with perforation. One hundred cases were studied and anaerobic bacteria were cultivated in 96 of these; at least 18 different species of anaerobes were recovered. *Bacteroides melaninogenicus* was recovered from 89 cases, anaerobic streptococci from 64, anaerobic cocci from 27, and Gram-negative anaerobic bacilli from 21. In the report of Wilson *et al.* (1972), on anaerobic bacteremia, 13 per cent of the 52 patients with anaerobic Gram-negative bacillary bacteremia had peritonitis as the portal of entry. Other papers confirm the major role of anaerobes in infections of the peritoneum (Sandusky *et al.*, 1942; Wright *et al.*, 1949; Gillespie & Guy, 1956; Schatten, 1956; Beazley *et al.*, 1972).

INFECTIONS OF THE LIVER

Considerable evidence is accumulating to indicate that anaerobic bacteria are a major cause of liver abscesses, if not the predominant cause. In the series of liver abscesses reported by Sabbaj *et al.* (1972), anaerobes were present in 45 per cent of cases. However, only a small percentage of the specimens were cultured in the research laboratory which had a much higher recovery rate for anaerobes than the clinical laboratory. It seems likely that anaerobes were responsible for well over half of the cases of liver abscess in that series. These authors also reviewed several major series of liver abscesses and found that, among 249 cases in 12 different series, 22 per cent of the cultures were sterile. The clinical features of anaerobic liver abscess are not different from those of pyogenic liver abscess in general, in keeping with the conclusion that most pyogenic liver abscesses are caused by anaerobic bacteria. Table 8.1 summarizes the isolations of anaerobic organisms from pyogenic liver abscess in a survey of the literature.

Blood cultures, usually reported as negative in liver abscess, were positive in 54 per cent of the cases described by Sabbaj *et al.* (1972). Recovery of anaerobes on blood

Table 8.1. Anaerobic organisms recovered from liver abscess (Summary of literature covering 229 patients)

Organism	No. Isolates
Anaerobic streptococci	42
Microaerophilic streptococci	19
Anaerobic Gram-positive cocci	6
Anaerobic Gram-negative cocci	1
Bacteroides species	12
Bacteroides fragilis	32
Bacteroides oralis	2
Bacteroides serpens	3
Bacteroides melaninogenicus	7
Fusobacterium species	33
Fusobacterium girans	2
Fusobacterium necrophorum	47
Unidentified anaerobic Gram-negative rods	16
Clostridium	29
Actinomyces	27
Eubacterium species	2
Bifidobacterium	1
Propionibacterium	1
Bacterium halosepticum	1
Unidentified anaerobic Gram-positive rods	5
'Anaerobe'	3
Total	291 strains

culture in the absence of an obvious portal of entry may be an important clue to liver abscess. Fulginiti *et al.* (1968) recorded 5 infections in liver transplant patients due to *Bacteroides fragilis*. Each of these patients had bacteraemia and one each also had subphrenic and hip abscesses and a septic infarction of the liver.

INFECTIONS OF THE GALLBLADDER AND BILIARY TREE

Clostridium perfringens infections of the biliary tract have been described more frequently than infection due to non-sporing anaerobes. Nevertheless, there are a number of reports of cases of cholecystitis and of cholangitis involving, particularly, *Bacteroides fragilis*. It is likely that a prospective study with good technique used for transport and culture of specimens would reveal a more important role for non-sporeforming anaerobes than has been recognised to date.

APPENDICEAL INFECTIONS

In the series of Bornstein *et al.* (1964), a quarter of the *Bacteroides* infections in a large series of anaerobic infections of all types involved the appendix. In the study of Felner & Dowell (1971), involving 250 cases of bacteremia due to Gram-negative anaerobic bacilli, over 10 per cent of the cases originated with appendiceal lesions. The excellent study of Altemeier (1938) on the bacterial flora of peritonitis associated with acute perforated appendicitis has already been described above. The results of early studies are impressive. Veillon & Zuber (1898) studied 22 cases of appendicitis: one yielded only aerobic bacteria, 2 yielded only anaerobic bacteria, and the rest had anaerobes as the predominant flora, with only rare streptococci and *E. coli*. The most common anaerobe found was *Bacteroides fragilis*. Heyde (1911) reported on 102 cases of appendicitis from which he recovered a total of 274 anaerobic and 153 aerobic strains, while Brütt (1923) reported the recovery of 45 anaerobic streptococci from 107 cases of appendicitis. Weinberg *et al.* (1928) reported on 150 patients with appendicitis, from whom they recovered Gram-negative anaerobes 60 times, anaerobic cocci and streptococci 40 times, and a total of 107 other anaerobic strains.

DIVERTICULITIS

It seems to be not generally appreciated that anaerobes are almost always involved in infections associated with diverticulitis. The significance of such infections is underlined by the frequency with which they predispose to anaerobic bacteraemia. Felner & Dowell (1971) found that in their series of 250 cases of *Bacteroides* bacteraemia, 20 originated from diverticulitis, and, in the series of Wilson *et al.* (1972), diverticulitis with perforation accounted for 6 per cent of bacteremias due to anaerobic Gram-negative bacilli.

PANCREATIC INFECTIONS

The fact that half of pancreatic abscesses grow multiple organisms, with a heavy predominance of enteric species, suggests that direct transmural penetration of the adjacent transverse colon may commonly occur (Warshaw, 1972). It is surprising, therefore, that anaerobes have not been recovered more frequently from pancreatic abscesses. This may reflect inadequate collection, transport, and anaerobic culture techniques. For example, Steedman *et al.* (1967) found that 2 of 13 cases of pancreatic abscess showed no growth, even though frank pus was cultured. There are, however, several reports of anaerobic infection in association with pancreatitis, pancreatic abscess, and pancreatic carcinoma (Altemeier & Alexander, 1963; Bodner *et al.*, 1970; Felner & Dowell, 1971; Norris, 1901; Serrano-Rios *et al.*, 1969 and Siler & Wulsin, 1950).

SPLENIC INFECTIONS

Anaerobic bacteria occasionally are involved in abscess of the spleen.

INFECTIONS RELATED TO THE STOMACH AND DUODENUM

Harvey *et al.* (1957) reported 3 cases of primary actinomycosis originating in perforated gastric ulcers, as well as 3 cases of secondary involvement of the wall of the stomach in actinomycosis. Other anaerobes are more often involved. Weiss (1943) recovered *Bacteroides melaninogenicus* from 2 cases of abdominal infection secondary to perforated duodenal ulcer and Wright *et al.* (1949) reported 16 cases of peritonitis due to ruptured gastro-duodenal ulcers, five of which yielded anaerobes. Schatten (1956) reported 3 cases of anaerobic peritonitis following perforated gastric or duodenal ulcers with surgical repair and Felner & Dowell

(1971) reported 3 cases of anaerobic bacteraemia secondary to ruptured gastric or duodenal ulcers, of which 2 were due to *Bacteroides fragilis*.

There are several other individual case reports involving infection with *Bacteroides*, *Fusobacterium*, *Actinomyces* and other anaerobes.

PERINEAL, PERI-RECTAL AND SIMILAR INFECTIONS

Anaerobic bacteria are very commonly involved in these infections. Bornstein *et al.* (1964) reported a perineal wound infection due to anaerobic streptococci. A few other case reports describe infections of these types involving Gram-negative anaerobic bacilli. Actinomycosis has been reported several times as a cause of perineal lesions. Ohnishi *et al.* (1972) recovered anaerobes (*Bacteroides* plus two *Peptococcus* strains) from three of 17 cases of perineal abscess.

FEMALE GENITAL TRACT INFECTIONS

Anaerobic bacteria are involved in a wide variety of types of infection of the female genital tract as indicated in Table 8.2. Conditions predisposing to or associated with such infections are listed in Table 8.3. The frequency with which anaerobic bacteria are involved in various obstetric and gynecological infections varies from one series to another, reflecting to some extent differences in patient material but probably primarily related to differences in effectiveness of transport systems and anaerobic culture techniques. The major infecting organisms, both anaerobic and aerobic, are noted in Table 8.4. Felner & Dowell (1971) found that 41 of 250 cases of *Bacteroides* bacteraemia originated from the genitourinary tract, half of these associated with postpartum or postabortion complications and some additional cases related to cervical or uterine tumours. In the series reported by Pearson & Anderson (1967), 40 of 76 such bacteremias were of female genital tract origin (incomplete septic abortion and puerperal infection). In a series of 173 cases of anaerobic Gram-negative bacillary bacteraemia encountered in a review of the literature from 1898 to 1956, Gunn (1956) found 17 of female genital tract origin, whereas Love (1972), reviewing the literature on such bacteraemias from 1956 to 1971, found that 116 of 458 cases had the genital tract as the point of origin.

Table 8.2. Types of anaerobic infection of the female genital tract.

Endometritis, pyometra
Myometritis
Parametritis
Pelvic cellulitis, abscess
Pelvic thrombophlebitis
Peritonitis, intra-abdominal abscess
Bacteraemia, with or without shock or intravascular haemolysis
Metastatic infection - particularly of lung, liver, brain, heart valves, kidneys
Wound infection or abscess
Vulval abscess
Bartholinitis, Bartholin abscess
Skenitis, Skene's abscess
Paraclitoridal abscess
Vaginitis, abscess of vaginal wall, paravaginal abscess
Salpingitis, tubal abscess
Ovarian or tubo-ovarian abscess
Abscesses of adjacent parts - groin, perirectal or ischiorectal areas, paraurethral, leg, abdominal wall
Chorioamnionitis
Fetal emphysema
Intrauterine or neonatal pneumonia, sepsis

Table 8.3. Predisposing or associated conditions in patients with anaerobic infections of the female genital tract.

Pregnancy
Puerperium, particularly with premature rupture of membranes, prolonged labour, or postpartum haemorrhage
Abortion - spontaneous or induced
Malignancy
Irradiation
Obstetric or gynecological surgery
Cervical cauterization
Endocervical or vaginal stenosis
Uterine fibroids
Secondary infection of old gonococcal salpingitis
Intra-uterine contraceptive devices

Table 8.4. Organisms involved in female genital tract infections.

I. Anaerobes

Anaerobic Gram-positive streptococci or cocci

Microaerophilic Gram-positive streptococci or cocci

Veillonella

Bacteroides fragilis

Bacteroides melaninogenicus

Other *Bacteroides* species

Fusobacterium (Sphaerophorus) necrophorum

Other *Fusobacterium* species

Clostridium perfringens

Actinomyces israelii

II. Facultative organisms

Escherichia coli

Other *Enterobacteriaceae*

Group A beta-hemolytic streptococci

Enterococci

Other streptococci

Staphylococcus aureus

Gonococcus

Haemophilus vaginalis

Proteus

Pseudomonas

In a series of 100 cultures which yielded anaerobic Gram-negative bacilli, Smith *et al.* (1968) found 48 isolates of these organisms from the female genital tract, and, in a similar series comprising 690 cultures, Lodenkamper & Stienen (1955) found that at least 83 of these cultures were from the female genital tract. Ueno (1973) recovered anaerobes in pure culture from 15 specimens of pus among 86 from gynecological patients and from 3 of 12 specimens of uterine discharge. Carter (1963) stated that non-sporeforming anaerobic bacilli are the most common isolates he had encountered from cases of Bartholinitis, Skenitis, pyometra, and postpartum and post-abortion endometritis. He stated that this is also true in the case of pelvic inflammatory disease, if one eliminates cases due to the gonococcus and tubercle bacilli.

Ledger *et al.* (1971) note that when *Bacteroides* species are isolated from infections in obstetric and gynecological patients, that convalescence was typically prolonged.

Robinson (1961) stated that streptococci, usually anaerobic or microaerophilic, were the commonest organisms cultured from pelvic abscesses (both cul-de-sac and tubo-ovarian). He also noted that these organisms were the rule in cases of puerperal or postabortion origin.

POSTABORTION SEPSIS

The excellent study of Rotheram & Schick (1969) involved detailed bacteriological studies of both blood and cervical cultures. From 69 cervical cultures (six cultured only aerobically), 129 anaerobic isolates were recovered. Anaerobes were found in abundance 67 times whereas this was true for aerobes only 21 times. The analysis of their 22 cultures which showed light or no anaerobic growth is also of interest. In this group of 22 cultures, there were four with an abundant growth of microaerophilic organisms, 11 with scanty growth of any kind, and one which was overgrown with *Proteus*, the latter showing many Gram-positive cocci on smear (presumably anaerobic). There were only six cultures in which there was moderate or abundant growth of *E. coli*.

The presentation of patients with septic abortion is usually impressive with high fever, chills, leukocytosis, uterine tenderness and foul-smelling cervical discharge which may be purulent and hemorrhagic. Nevertheless, most of these patients recover readily following evacuation of the uterus.

The complications of septic abortion, however, may be severe. Bacteria may migrate through the uterine wall by way of lymphatic channels and small venules to produce inflammation and abscesses in the broad ligament and the retroperitoneal spaces of the pelvis. Such direct extension may result in tubal or tubo-ovarian abscesses which may rupture to produce local or remote intra-abdominal or pelvic abscess or peritonitis. Septic thrombophlebitis of the pelvic and ovarian veins is common and should be presumed to be present in patients who are not responding to what should be appropriate antimicrobial therapy, particularly if there is no evidence of abscess formation or other obvious complication. This thrombophlebitis accounts for continued septicemia and may lead to metastatic infection of the lungs, brain, liver, kidney and heart valves, although extensive spread of this type is uncommon in the antimicrobial era.

PUERPERAL SEPSIS

Anaerobes also play an important role in puerperal sepsis. Anaerobic cocci predominate but *Bacteroides* and other Gram-negative anaerobic bacilli, clostridia, and other anaerobes may also be involved.

There are several reports of puerperal infection in association with Caesarean section. (Harris & Brown, 1927; Ledger *et al.*, 1971; Tracy *et al.*, 1972).

One important reason for the lesser virulence of aerobic pathogens which may be incolved in puerperal sepsis is the fact that they seldom produce thrombophlebitis, in contrast with anaerobic Gram-negative bacilli and anaerobic streptococci. The clinical features of puerperal sepsis depend on the extent of the infection. With endometritis alone, there is usually only mild fever and some increase in lochia for the first several postpartum days. With spread of the infection, there is higher fever, chills, lower abdominal pain and specific signs of involvement of the peritoneum and/or other structures. Putrid odour to the lochia is typical of anaerobic endometritis and the unique morphology of the anaerobes involved may often guide the clinician to proper presumptive etiologic diagnosis.

The important conditions predisposing to puerperal sepsis are premature rupture of membrances, prolonged labour and postpartum hemorrhage. Antimicrobial therapy is usually adequate, but heparin may be helpful in the case of thrombophlebitis and, on occasion, it may be necessary to ligate the inferior vena cava when recurrent septic embolization occurs despite anticoagulation.

OTHER UTERINE INFECTIONS

Pyometra is a relatively frequent infection involving anaerobic bacteria. It is seen particularly in association with malignancy and especially when the lesion has been irradiated, but it may follow cervical cauterization or may be a consequence of endocervical stenosis. It is a benign condition, and the symptoms include fever, local pain, bleeding and discharge of foul-smelling pus which may contain gas bubbles. Treatment with drainage and antimicrobial compounds is usually effective. Again, the results vary considerably from one series to another, but it is clear that anaerobes may play a very important role in this type of infection. Sugiyama *et al.* (1972) found that 39 of 46 cultures from patients with infection in association with carcinoma of the cervix contained anaerobic bacteria, 23 of these yielding anaerobes in the absence of any aerobes.

TUBO-OVARIAN INFECTIONS

The pattern is again similar to that observed in other female genital tract infections, some series showing a very high incidence of anaerobes and others showing a considerably lower incidence. In this connection, it is important to note that Altemeier (1940) reviewed 31 series in the literature prior to the time of his publication and noted that among a total of 1,179 cases, 53.6 per cent were sterile on culture. Anaerobic cocci and streptococci predominate in tubo-ovarian abscesses, but anaerobic Gram-negative bacilli are also encountered relatively frequently. Other anaerobes, including clostridia, are relatively uncommon.

Tubo-ovarian and other pelvic abscesses may be genital or nongenital in origin. Although infection may be haematogenous in origin, the majority arise by contiguity from septic abortion, uterine perforation, appendicitis, diverticulitis, rupture of any abdominal viscus, or following abdominal or pelvic surgery. Other background factors include trauma, abdominal or uterine malignancy and degenerating fibroids.

Purulent vaginal discharge is relatively uncommon and less than half of the patients have a history of prior pelvic infection. Culdocentesis is very important in establishing the diagnosis.

With regard to therapy, aggressive surgical management is important. Immediate surgery is crucial in the event of ruptured tubo-ovarian abscess. At the time of surgery, it is important to explore the entire abdomen carefully, looking for concealed pockets of pus. In young patients, it is desirable to remove only the diseased organs. However, the nature of the disease is such that total abdominal hysterectomy and bilateral salpingo-oophorectomy is frequently necessary. Prolonged antimicrobial therapy is very important, as it is in virtually all anaerobic infections, because of a marked tendency for relapse. Patients should be treated for at least 2-3 weeks after all evidence of infection has disappeared.

BARTHOLIN GLAND INFECTION AND OTHER SUPERFICIAL INFECTIONS

Anaerobic bacteria are commonly involved in Bartholinitis. In this case, the anaerobic non-sporeforming Gram-negative bacilli are more commonly encountered than are anaerobic or microaerophilic cocci. Skene's gland infection may also involve anaerobes. Other lesser infections involving anaerobes include noma, abscess of the external genitalia, vaginal abscess, infected vaginal cyst, and infected condylomata.

The difficulty in establishing the significance of anaerobic bacteria in superficial infections of the vulva and vagina, where there are large numbers of anaerobes normally as indigenous flora, can well be appreciated. Nevertheless, certain features of these infections, such as the foul odour to discharges or lesions and the tissue necrosis, as well as increased numbers of anaerobes in the lesions, and concurrent bacteraemia with anaerobes, may make the distinction possible.

POST-OPERATIVE GYNECOLOGICAL INFECTIONS

Anaerobic bacteria of all types are frequently involved in postoperative infections following various types of gynecological surgery. As indicated earlier, gynecological infections may result from other types of surgery as well. Cases of infection following Caesarean section were referred to earlier.

An excellent discussion of post-operative pelvic infections is presented by Ledger (1969). Surprisingly, young women seem to have the highest incidence of infection following surgery, probably primarily because of the vaginal surgical approach. This is, of course, in turn related to the abundant bacterial flora of the vagina. A thorough surgical preparation of the vagina in the operating room, using an iodophor solution, is important in reducing the number of contaminating organisms. If the interval between cone biopsy of the cervix and hysterectomy is too long, bacteria will proliferate in the devitalized biopsy site where poor tissue perfusion and clotted blood provide an excellent environment for growth of anaerobes and other bacteria. When feasible, the surgery should be performed within two days of the time of biopsy or not until six weeks or more later, to allow adequate healing. The management of the ovarian pedicle may also be an important factor increasing the morbidity after vaginal hysterectomy. The close proximity of the ovarian pedicle to the infected post-operative vaginal cuff permits the spread of organisms to the adnexa and may contribute to later development of adnexal abscess.

When Caesarean section is carried out after prolonged rupture of the membranes, there is a distinct hazard that amnionitis will be present and that there will be excessive contamination of the operative field with endogenous bacteria, including anaerobes. Aggressive obstetric management should minimize delay from the onset of labour and thus make bacterial contamination of the operative field less likely. A low segment Caesarean section will usually minimize or prevent spillage of amniotic fluid or infected material from the uterus into the peritoneal cavity. When it appears that the operative field is likely to be contaminated, aerobic and anaerobic culture should be obtained at the time of operation so that in the event of subsequent infection, intelligent therapy may be applied much sooner. In the event of overt infection, of course, therapy should be started without waiting for the culture report, based on the likely infecting organism and examination of the Gram-stained smear.

Post-operative infection is manifested either as a localized vaginal cuff abscess, a diffuse pelvic cellulitis, or adnexal infection. Distinction between these different entities is not always clear cut and, not infrequently, post-operative infection will be manifested simply by fever. Nevertheless, management will be assisted by the precise assignment of a specific diagnostic category, whenever possible.

MISCELLANEOUS

Septic thrombophlebitis should be suspected in women in the puerperal period, or in the period following gynecological surgery, when there is unexplained fever, with or without vague abdominal symptoms and particularly where there are no physical findings and there has not been a response to antimicrobial agents. Such thrombophlebitis is commonly encountered in anaerobic infection. Blood cultures may be positive, particularly if appropriate technique is used to recover anaerobic organisms, but they may, not uncommonly, be negative. Pelvic thrombophlebitis may be difficult to diagnose and is basically a diagnosis of exclusion. Addition of heparin to the regimen will often bring about response and may actually serve as a diagnostic test.

Anaerobic bacteria are often involved in septic shock in obstetric and gynecological patients. While the endotoxin of certain Gram-negative anaerobic bacilli may be involved in some of these cases, it is clear that there are other mechanisms as well, as is well known with aerobic bacteraemia.

Abscesses of adjacent parts such as the groin, perirectal or ischiorectal areas, the upper leg, and the abdominal wall may be related to anaerobic infections of female genital organs.

URINARY TRACT INFECTION

A number of uncritical reports on anaerobic urinary tract infection, which failed to take into consideration the presence of anaerobes as normal flora on the urethral mucosa (Finegold *et al.*, 1965) have led to confusion. Nevertheless, it is well documented that a variety of urinary tract infections may involve anaerobic organisms. Table 8.5 lists these infections.

With renal infection, there is a distinct association with stones, primarily related to tissue necrosis adjacent to the stone. Other factors predisposing to anaerobic urinary tract infection are malignancy, obstruction, and previous renal tuberculosis. Distinctive features that may be seen in anaerobic urinary tract infection, as in other types of anaerobic infection, include tissue necrosis, gas in tissues and foul odour.

Table 8.5. Urinary tract infections involving anaerobes.

Peri-urethral cellulitis, abscess, or gangrene
Cowperitis
Chronic prostatitis
Prostatic abscess
Cystitis
Pyelocystitis
Pyelonephritis
Pyonephrosis
Renal abscess
Perirenal abscess
Testicular abscess
Wound infection following nephrectomy, nephrostomy or other open surgery
Actinomycosis - usually a chronic, circumscribed parenchymal lesion. May progress to pyonephrosis or perinephric abscess.

Data presented by Segura *et al.* (1972) indicate an incidence of anaerobic bacteriuria of at least 1.3 per cent in patients with significant bacteriuria. In this study, patients who seemed to have significant bacteriuria on the basis of a Gram-stain of the urine, and who had negative aerobic cultures, were subjected to suprapubic bladder aspiration and anaerobic culture. Ten of 17 patients so studied had significant anaerobic bacteruria, usually with aerobic or facultative organisms present as well. All of these patients had received multiple courses of antimicrobial therapy for urinary tract infection characterized by resistance to treatment and frequent recurrence. There was a significant incidence of obstructive uropathy and other associated urological disorders.

INFECTION OF SKIN, SOFT TISSUE AND MUSCLE

Anaerobic infection of the skin and soft tissue includes such entities as cellulitis, pyoderma, infected sebaceous or inclusion cysts, hidradenitis suppurativa, infection of decubitus or other superficial ulcers, infected pilonidal cysts, cutaneous abscesses, anaerobic gas-forming cellulitis, bacterial synergistic gangrene, chronic undermining ulcer of Meleney, necrotizing fasciitis, infection following bites or other trauma, noma, and tropical ulcer.

Superficial ulcers secondary to vascular disease (arterial or venous) are often infected with anaerobic bacteria, particularly in diabetics. There are several reports of subcutaneous abscesses involving anaerobic bacteria in narcotic addicts who are 'skin-poppers'.

Anaerobic cellulitis is an acute infection of soft tissue. The term is not properly descriptive but persists because of common usage. It is synonymous with gas abscess. Many of the patients with this condition are diabetics. Although coliform bacilli and other gas-producing non-anaerobes may be involved, anaerobic bacteria predominate in this type of lesion. Differentiation from gas gangrene may require surgery which, in any case, is the definitive approach to therapy.

Bacterial synergistic gangrene is a painful, chronic gangrene of the skin which is usually a postoperative complication. Classically, it is due to a combination of a microaerophilic, non-haemolytic streptococcus, which is found primarily in the spreading periphery of the lesion, and *Staphylococcus aureus*, which is found primarily in the central gangrenous zone.

Chronic undermining ulcer or 'Meleney's ulcer' is a slowly progressive infection of the subcutaneous tissues associated with tenderness and ulceration of the overlying skin. The ulcer is seen most frequently following lymph node surgery in the neck, axilla or groin and after operations on the intestinal and genital tracts. Multiple ulcers and sinuses may develop at a distance from the original ulcer, with undermining of the intervening skin. Classically, the causative organism is a microaerophilic haemolytic streptococcus. However, anaerobic streptococci are found at times.

Necrotizing fasciitis, otherwise known as hospital gangrene, is a serious infection which commonly originates in musculoskeletal wounds but which may appear in an operative would or after injury. A pathognomonic feature is subcutaneous and fascial necrosis with undermining of the skin. This disease has been thought to be caused only by pyogenic aerobic cocci, but recently there have been a number of reports of single cases in which anaerobic bacteria have been involved.

Fournier's gangrene usually starts in or around the scrotum and behaves like anaerobic cellulitis or necrotizing fasciitis. This disease may remain limited to the scrotum or may also extend to the penis, the abdominal wall and the perineum.

Among the most impressive anaerobic infections secondary to trauma are those due to bites, particularly human bites. The anaerobes most often involved are anaerobic or micro-aerophilic cocci, *Fusobacterium nucleatum* and *Bacteroides melaninogenicus*.

Tropical ulcer and noma are seen in tropical and sub-tropical regions and underdeveloped areas. These co-called fusospirochetal infections, like 'Vincent's angina', are probably mixed anaerobic infections in which *B. melaninogenicus* plays an important role (see Chapter 6).

Infected vascular gangrene is another condition in which gas-producing anaerobes are found frequently. In this situation the muscle, or the entire limb, has already died as a result of circulatory insufficiency and the bacteria act primarily as saprophytes.

On occasion, myositis or myonecrosis may be caused by anaerobes other than clostridia. Chief among these is the anaerobic streptococcus. Anaerobic streptococcal myonecrosis may resemble subacute clostridial gas gangrene. Gram-negative anaerobic bacilli, on occasion, may produce a low grade infection of muscle.

MISCELLANEOUS ANAEROBIC INFECTIONS

Venereal or genital fusospirochaetosis, also called the 'Sixth Venereal Disease', would perhaps better be called anaerobic balanoposthitis. The principal complaints are pain and foul discharge. Shallow ulcerations may appear on the glans and the prepuce. There may also be secondary infection of chancroid, syphilitic lesions, papillomas, and condylomata acuminata with anaerobic bacteria.

Abscesses of the buttock, in addition to those involving the perirectal or ischiorectal spaces, often involve anaerobic bacteria. Also non-puerperal breast abscesses, particularly those that are subareolar and recurrent, may often involve anaerobes, especially Gram-negative anaerobic bacilli.

Suppurative lymphadenitis may be due to anaerobes. *Propionibacterium*, a common skin 'contaminant', may produce this type of infection.

THERAPY OF ANAEROBIC INFECTIONS

Surgical aspects of therapy are extremely important. Drainage of collections of pus and excision of necrotic tissue are commonly necessary and in minor infections may be all that is required. Obstructions must be relieved. Septic phlebitis occurs in some patients and if pulmonary emboli result, anticoagulant therapy or venous ligation may be indicated. Treatment of shock and disseminated intravascular coagulation, when they occur, and general supportive measures are important aspects of therapy.

There are a number of good drugs available for treatment of anaerobic infections. However, individual groups or strains of anaerobes may be resistant to various agents so that care is necessary in prescribing therapy. Other factors such as toxicity, impact on the indigenous flora, pharmacologic properties of the drug, allergy in the patient, available dosage forms, bactericidal activity and the renal and hepatic functional status of the patient, help to determine the choice of therapy.

Penicillin is active against most anaerobes other than *Bacteroides fragilis*, but *Bacteroides fragilis* is by far the most common anaerobe isolated from infectious processes in man. Certain strains of anaerobic cocci and of *B. melaninogenicus* also require as much as 8.0 μg/ml for inhibition, and occasional strains of *Fusobacterium varium* are resistant. For therapy of serious infections or infections due to relatively resistant strains, intravenous doses of 25-30 million units daily (for adults) are desirable. For other infections, a daily dosage of 2 to 10 million units is adequate. After the acute infection has responded, certain infections (such as most cases of lung abscess) requiring prolonged therapy may be managed successfully with oral benzylpenicillin (400,000-800,000 units four times daily). Ampicillin and cephaloridine are usually comparable to benzylpenicillin but other penicillins and cephalosporins are often less active. Certain agents, such as carbenicillin, may be valuable because very high blood levels are achieved safely.

Tetracycline is less useful than previously because many anaerobes have developed resistance to it (Sutter *et al.*, 1972, and see Chapter 4). Dosage (adult) is 500 mg I.V. 12-hourly orally or 500 mg 6-hourly orally. Intravenous therapy gives distinctly higher blood levels. Two new derivatives, doxycycline and minocycline, are much more active. Dosage (adult) of doxycycline is 100 to 200 mg per day intravenously, given once daily in 400 ml volume, over a 2-4 hour period. The oral dose is 100 mg every 12 hours the first day, then 100 mg in a single dose daily. With minocycline, the dose (i.v. or orally) is 100 mg twice daily following an initial dose of 200 mg.

Chloramphenicol is active against all anaerobes, with only rare exceptions, and is very effective clinically. It penetrates the central nervous system well. However, because of its serious and unpredictable toxicity, it should be reserved for seriously ill patients and used only until bacteriological data indicate that a less toxic drug would be suitable. Nevertheless, I would regard chloramphenicol as the drug of choice for a suspected very serious anaerobic infection of uncertain etiology. Initial dosage in seriously ill adults is 1.0 g 6-hourly i.v.; subsequently, the dosage may be reduced to 500 mg 6-hourly i.v. or orally.

Clindamycin has broad activity against anaerobes, including *Bacteroides fragilis*. Only *Fusobacterium varium*, among the non-sporeforming anaerobes, commonly shows resistance to this agent. However, a number of strains of *Clostridium*, particularly among species other than *C. perfringens*, are not uncommonly resistant. It is well established as being effective clinically (Bartlett *et al.*, 1972). Necrotizing colitis (pseudomembranous enterocolitis) has been described with this agent, but is apparently very uncommon. The dosage (adults) is 150-300 mg every 6 hours orally and 600 mg to 2700 mg per day (in two to four equal doses) parenterally. Daily doses of as much as 4.8 grams have been given intravenously in critically ill patients.

Metronidazole is active only against anaerobic microorganisms, but certain of the cocci, particularly those that are microaerophilic, and *Actinomyces* are resistant. It is consistently bactericidal versus *Bacteroides fragilis*.

Early clinical experience is promising (Tally *et al.*, 1972). A parenteral form of the drug will become available for trial in the near future. Dosage regimens remain to be established. However, we have successfully used adult dosages of 250 mg to 750 mg three times daily.

Therapy with antimicrobial agents in anaerobic infections must be intensive and prolonged. These infections have a considerable tendency to relapse.

REFERENCES

Altemeier, W.A. (1940). The Anaerobic Streptococci in Tubo-Ovarian Abscess. *American Journal of Obstetrics*. 39, 1038.

Altemeier, W.A. (1938). The Bacterial Flora of Acute Perforated Appendicitis with Peritonitis. *Annals of Surgery*, 107, 517.

Altemeier, W.A. & Alexander, J.W. (1963). Pancreatic Abscess - A Study of 32 Cases. *Archives of Surgery*, 87, 96.

Altemeier, W.A., Culbertson, W.R., Fullen, W.D. & Shook, C.D. (1973). Intra-abdominal Abscesses. *American Journal of Surgery*, 125, 70.

Armfield, A.Y., Felner, J.M., Thompson, F.S., McCroskey, L.M. & Balows, A. (1971). Comparison of Clinical and Bacteriologic Data from *Eubacterium filamentosum* and *Clostridium innocuum*. (M266). *Bacteriological Proceedings - 1971*, 109.

Azar, H. & Drapanas, T. (1968). Relationship of Antibiotics to Wound Infection and Enterocolitis in Colon Surgery. *American Journal of Surgery*, 115, 209.

Bartlett, J.G., Sutter, V.L. & Finegold, S.M. (1972). Treatment of Anaerobic Infections with Lincomycin and Clindamycin. *New England Journal of Medicine*, 287, 1006.

Beazley, R.M., Polakavetz, S.H. & Miller, R.M. (1972). Bacteroides Infections on a University Surgical Service. *Surgery, Gynecology and Obstetrics*, 135, 742.

Bernard, H.R. & Cole, W.R. (1963). Wound Infections Following Potentially Contaminated Operations. *Journal of the American Medical Association*, 184, 290.

Bodner, S.J., Koenig, G. & Goodman, J.S. (1970). Bacteremic Bacteroides Infections. *Annals of Internal Medicine*, 73, 537.

Bornstein, D.L., Weinberg, A.N., Swartz, M.N. & Kunz, L.J. (1964). Anaerobic Infections-Review of Current Experience. *Medicine*, 43, 207.

Brütt, H. (1923). Die Bedeutung der Anaeroben Streptokokken für die Destruktive Appendicitis. *Beitr. Klin. Chir.*, 129, 175.

Carter, B. (1963). Anaerobic Infections in Obstetrics and Gynaecology. *Proceedings of the Royal Society of Medicine*, 56, 1095.

Felner, J.M. & Dowell, V.R., Jr. (1971). 'Bacteroides' Bacteremia. *American Journal of Medicine*, 50, 787.

Finegold, S.M. (1969). Intestinal Bacteria. *California Medicine*, 110, 455.

Finegold, S.M., Marsh, V.H. & Bartlett, J.G. (1971). Anaerobic Infections in the Compromised Host. *Proceedings International Conference on Nosocomial Infections, Atlanta, American Hospital Association*, Chicago, p.123.

Finegold, S.M., Miller, L.G., Merrill, S.L. & Posnick, D.J. (1965). Significance of Anaerobic and Capnophilic Bacteria Isolated from the Urinary Tract. *Progress in Pyelonephritis*, Ed. Kass, E.H. F.A. Davis Co., Philadelphia, p.159.

Fulginiti, V.A., Scribner, R., Groth, C.G., Putnam, C.W., Brettschneider, L., Gilbert, S., Porter, K.A. & Starzl, T.E. (1968). Infections in Recipients of Liver Homografts. *New England Journal of Medicine*, 279, 619.

Furuta, K. & Tsuchiya, T. (1973). Abdominal Wall Abscesses due to Bacteroides fragilis. *Journal of the Japanese Association for Infectious Diseases*, 47, 44.

Gillespie, W.A. & Guy, Joan. (1956). Bacteroides in Intra-Abdominal Sepsis. Their Sensitivity to Antibiotics. *Lancet*, 1, 1039.

Gorbach, S.L., Thadepalli, H. & Norsen, J. Anaerobic Micro-organisms in Intraabdominal Infections. In *Anaerobic Bacteria: role in disease*. Ed. Balows, A., De Haan, R.M., Dowell, V.R. Jr. & Guze, L.B. Charles C. Thomas, Springfield, Illinois. p.399.

Gorbach, S.L., Thadepalli, H. & Norsen, J. (1972). Isolation of 25 Clostridia Species from Clinical Sources. Presented at 12th Interscience Conference on Antimicrobial Agents and Chemotherapy, Atlantic City, New Jersey.

Gunn, A.A. (1956). Bacteroides septicaemia. *Journal of the Royal College of Surgeons, Edinburgh*, 2, 41.

Harris, J.W. & Brown, J.H. (1927). Description of a New Organism that may be a Factor in the Causation of Puerperal Infection. *Bulletin of the Johns Hopkins Hospital*, 40, 203.

Harvey, J.C., Cantrell, J.R. & Fisher, A.M. (1957). Actinomycosis: Its Recognition and Treatment. *Annals of Internal Medicine*, 46, 868.

Heinrich, S. & Pulverer, G. (1960). Über den Nachweis des Bacteroides melaninogenicus in Krankheitsprozessen bei Mensch und Tier. *Zeitschr. f. Hyg.*, 146, 331.

Heyde, M. (1911). Bakteriologische und experimentelle Untersuchungen zur Aetiologie der Wurmfortsatzentzundung (mit besonderer Berucksichtigung der anaeroben Bakterien). *Beitr. Klin. Chir.*, 76, 1.

Hoffman, K. & Gierhake, F.W. (1969). Postoperative Infection of Wounds by Anaerobes. *German Medical Monthly*, 14, 31.

Kuklinca, Arlington G. & Gavan, Thomas L. (1971). Anaerobic bacteria in postmortem blood cultures. Correlation with Lesions of the Gastro-intestinal Tract. *Cleveland Clinic Quarterly*, 38, 5.

Ledger, W.J. (1969). Postoperative Pelvic Infections. *Clinical Obstetrics and Gynaecology*, 12, 265.

Ledger, W.J., Sweet, R.L. & Headington, J.T. (1971). Bacteroides Species as a Cause of Severe Infections in Obstetric and Gynaecologic Patients. *Surgery, Gynaecology and Obstetrics*, 133, 837.

Lodenkämper, H. & Stienen, G. (1955). Importance and Therapy of Anaerobic Infections. *Antibiotic Medicine*, 1, 653.

Lodenkämper, H. & Stienen, G. (1956). The Treatment of Anaerobic Infections. *German Medical Monthly*, 1, 233.

Love, W.C. (1972). Clinical Aspects of Bacteroides Infections. In *Host Resistance to Commensal Bacteria. The Response to Dental Plaque*. Ed. Macphee, T. Edinburgh: Churchill Livingstone.

Magilligan, D.J., Jr. (1968). Suprahepatic Abscess. *Archives of Surgery*, 96, 14.

Marcoux, J.A., Zabransky, R.J., Washington, J.A. II, Wellman, W.E. & Martin, W.J. (1970). Bacteroides Bacteremia. *Minnesota Medicine*, 53, 1169.

Norris, C. (1901). Suppurative Pylephlebitis Associated with Anaerobic Micro-Organisms. *Journal of Medical Research*, 6, 97.

Ohnishi, T., Kanamori, M., Miyaji, H., Otani, H., Kato, T., Shinagawa, N., Tsuruga, N. & Shibata, S. (1972). Anaerobic Isolations from Surgical Infections; Clinical and Bacteriological Characteristics. In: *2nd Symposium on Anaerobic Bacteria and Their Infectious Diseases*, Eds and promotors - Ishiyama, S., Ueda, T., Kozakai, N., Suzuki, S. & Ueno, K. p.89.

Pearson, H.E. & Anderson, Gail V. (1967). Perinatal Deaths Associated with Bacteroides Infections. *Obstetrics and Gynaecology*, 30, 486.

Robinson, S.C. (1961). Pelvic Abscess. *American Journal of Obstetrics and Gynecology*, 81, 250.
Rotheram, E.B., Jr. & Schick, S.F. (1969). Nonclostridial Anaerobic Bacteria in Septic Abortion. *American Journal of Medicine*, 46, 80.
Sabbaj, J., Sutter, V.L. & Finegold, S.M. (1972). Anaerobic Pyogenic Liver Abscess. *Annals of Internal Medicine*, 77, 629.
Saksena, D.S., Block, M.A., McHenry, M.C. & Truant, J.P. (1968). Bacteroidaceae: Anaerobic Organisms Encountered in Surgical Infections. *Surgery*, 63, 261.
Sandusky, W.R., Pulaski, E.J., Johnson, B.A. & Meleney, F.L. (1942). The Anaerobic Nonhemolytic Streptococci in Surgical Infections on a General Surgical Service. *Surgery, Gynecology and Obstetrics*, 75, 145.
Schatten, W.E. (1956). Intraperitoneal Antibiotic Administration in the Treatment of Acute Bacterial Peritonitis. *Surgery, Gynecology and Obstetrics*, 102, 339.
Segura, J.W., Kelalis, P.P., Martin, W.J. & Smith, L.H. (1972). Anaerobic Bacteria in the Urinary Tract. *Mayo Clinic Proceedings*, 47, 30.
Serrano-Rios, M., Navarro, V., Fontan, J., Oliva, H. & Ramirez, J. (1969). Isolated Hepato-pancreatic Actinomycosis. Report of a Case Simulating an Acute Abdomen of Fatal Course. *Digestion*, 2, 262.
Shoemaker, M.V. (1960). Gram-Negative Anaerobic Bacilli: Isolation, Incidence, and Cultural Characteristics. *American Journal of Medical Technology*, 26, 279.
Siler, Vinton E. & Wulsin, J.H. (1950). Acute Pancreatitis. *Journal of the American Medical Association*, 2, 78.
Smith, C.D., Deane, C., Montgomery, J., Williams, R. & Sampson, C.C. (1968). Bacteroides. *Journal of the National Medical Association*, 60, 215.
Spaulding, E.H., Vargo, V., Michaelson, T.C., Vitagliano, R., Swenson, R.M. & Forsch, E. (1971). Value of a Prereduced Anaerobic Method in the Bacteriology of Abdominal Wounds. *Bacteriological Proceedings - 1971*, p.109.
Steedman, R.A., Doering, R. & Carter, R. (1967). Surgical Aspects of Pancreatic Abscess. *Surgery, Gynecology and Obstetrics*, 125, 757.
Stokes, E.J. (1958). Anaerobes in Routine Diagnostic Cultures. *Lancet*, 1, 668.
Sugiyama, Y., Yamaji, K. & Seiga, K. (1972). Anaerobic Bacteria Isolated from Exudates in Cancer of the Cervix. In *2nd Symposium on Anaerobic Bacteria and their Infectious Diseases (1972)*. Eds and promotors - Ishiyama, S., Ueda, T., Kozakai, N., Suzuki, S. & Ueno, K. p.125.

Sutter, V.L., Kwok, Y-Y. & Finegold, S.M. (1972). Standardized Antimicrobial Disc Susceptibility Testing of Anaerobic Bacteria. I. Susceptibility of Bacteroides fragilis to Tetracycline. *Applied Microbiology*, 23, 268.

Tally, F.P., Sutter, V.L. & Finegold, S.M. (1972). Metronidazole Versus Anaerobes. In Vitro Data and Initial Clinical Observations. *California Medicine*, 117, 22.

Thadepalli, H., Gorbach, S.L., Broido, P. & Norsen, J. (1972). A Prospective Study of Infections in Penetrating Abdominal Trauma. *American Journal of Clinical Nutrition*, 25, 1405.

Thadepalli, H., Gorbach, S.L., Broido, P.W., Norsen, J. & Nyhus, L. (1973). Abdominal Trauma, Anaerobes, and Antibiotics. *Surgery, Gynecology and Obstetrics*, 137, 270.

Thadepalli, H. & Gorbach, S. Clindamycin in Pure and Mixed Anaerobic Infections. *Archives of Internal Medicine* (In press).

Tracy, O., Gordon, A.M., Moran, F., Love, W.C. & McKenzie, P. (1972). Lincomycins in the Treatment of Bacteroides Infections. *British Medical Journal*, 1, 280.

Ueno, K. (1973). Personal Communication.

Veillon, A. & Zuber, A. (1898). Recherches Sur Quelques Microbes Strictement Anerobies et Leur Role en Pathologie. *Arch. de Méd. Exp. et Anat. Path.*, 10, 517.

Warshaw, A.L. (1972). Pancreatic Abscesses. *New England Journal of Medicine*, 287, 1234.

Weinberg, M., Prevot, A.R., Davesne, J. & Renard, C. (1928). Recherches Sur La Bactériologie et La Sérotherapie Des Appendicites Aigues. *Ann. Inst. Past.* (Supp.), 42, 1167.

Weiss, C. (1943). The pathogenicity of Bacteroides melaninogenicus and Its Importance in Surgical Infections. *Surgery*, 13, 683.

Wilson, W.R., Martin, W.J., Wilkowske, C.J. & Washington, J.A., II. (1972). Anaerobic Bactaeremia. *Mayo Clinic Proceedings*, 47, 639.

Wright, L.T., Metzger, W.I., Shapero, E.B., Carter, S.J., Schreiber, H. & Parker, J.W. (1949). Treatment of Acute Peritonitis with Aureomycin. *American Journal of Surgery*, 78, 15.

DISCUSSION

DR WILSON (BRUNEL UNIVERSITY): We have recently found that metronidazole may be a useful adjunct in radiotherapy.

Unfortunately, however, the doses that would be required would be of the order of 200 mg, not 3 times a day as usual, but per kg body weight. This would mean a dose of something like 14 g for a 70 kg man. Obviously, toxicity tests were necessary and we have been doing these on rats using a dose of 1 g/kilo. Some of the results are interesting because they may be of importance in chemotherapy. We find that metronidazole is rapidly degraded by the rat intestinal contents. In a 15 per cent anaerobic suspension we find that amounts of metronidazole of the order of 170 µg/ml are rapidly degraded with a half-life of 15 minutes. (This concentration is much higher than that normally active chemotherapeutically). If we give the rats benzylpenicillin for three days in the drinking water before looking at the activity of the caecal contents, we find that metronidazole is no longer rapidly removed. In view of Dr Finegold's paper showing metronidazole to be bactericidal, whereas in general other drugs are only bacteriostatic, this suggests that the combination of metronidazole and penicillin may, perhaps, be useful in chemotherapy. The second point is, that it does show that anaerobic caecal contents remove drugs very rapidly. In fact, it is possible that in the normal state, anaerobes such as *Bacteroides* rather than being pathogenic, have some sort of defensive role.

9. Infections due to Non-sporing Anaerobes: Some Illustrative Cases

A. T. WILLIS, S. E. J. YOUNG and I. R. FERGUSON

The non-sporing anaerobic bacteria, and especially the *Bacteroides*, are the commonest cause of anaerobic infections in man (Finegold *et al.*, 1972; Finegold & Rosenblatt, 1973; Levison, 1973). Such infections are usually of endogenous origin, because most of these organisms are obligate parasites that form part of the normal bacterial flora of the mucosal surfaces of the oropharynx, and the alimentary and genital tracts. From these situations the non-sporing anaerobes may invade adjacent tissues that are the seat of some pathological change. General predisposing factors to infection include debility, diabetes, malignant disease, anaemia, and blood loss. The trauma of abdominal surgery, especially bowel and gynaecological surgery, provides conditions favourable to the development of anaerobic infections, as does the physiological process of childbirth.

These organisms cause a variety of infections which are not commonly initiated in the vicinity of their normal habitats, where characteristically deep seated, foul smelling abscesses are produced. The subsequent development of septic thrombophlebitis explains some of the common features of these infections - local extension, bacteraemia, and metastatic abscess formation.

Like the *Staphylococcus*, these organisms produce a great diversity of infections, and many reports and reviews of infections referable to the three main portals of entry have been published over the last few years, especially in the American literature (see Chapters 7 and 8). The very common occurrence of non-clostridial anaerobic infection is not reflected in the United Kingdom literature of recent years. Apart from the earlier reports of human *Bacteroides* infections by Alston (1955), Gillespie & Guy (1956), Gunn (1956) and Stokes (1958), we can find only five British publications in which cases of *Bacteroides* infection are documented. Child *et al.* (1969) described a single case of *Bacteroides fragilis* endocarditis in a heroin addict, Marsden & Hyde (1971) described 6 cases of *Bacteroides corrodens* infection in children, Tracy *et al.* (1972)

presented 6 cases of *Bacteroides* infection - 4 following intestinal surgery, 1 following caesarean section, and 1 pulmonary infection, and Mitchell & Simpson (1973) described 5 cases of *Bacteroides* bacteraemia which developed after intestinal surgery. Okubadejo *et al.* (1973) referred to *Bacteroides* infections in 112 hospital patients; most of these infections were due to *Bacteroides fragilis*, and most of them were severe and were associated with major surgical procedures such as intestinal surgery for carcinoma and post-partum hysterectomy.

In the light of American experience it is remarkable that there is such a paucity of published clinical data on these exceedingly common infections in the United Kingdom. This is probably due partly to a general unawareness of the importance of the non-sporing anaerobes as causes of human disease, so that they are not considered clinically and not sought bacteriologically, and partly to poor liaison between the clinician and the microbiologist. Success in the isolation of anaerobes from clinical material is clearly dependent on the use of a good anaerobic technique. It is also greatly influenced by the type of specimen submitted for study, by the time interval between its collection and examination, and by the clinical information that accompanies it. These are matters which themselves depend, not only on a clinical awareness of the possibility of an anaerobic infection, but also on a constant dialogue between the clinician and the microbiologist; failure here is likely to result in early misdiagnoses and in unacceptable delays in establishing the correct diagnosis. We ourselves have not yet resolved these fundamental problems of communication, as is reflected in some of the case histories outlined below.

A clinical awareness of non-clostridial anaerobic infections should not be difficult to foster. Modern anaerobic equipment is reliable, readily available, and inexpensive, and anaerobic techniques are easy. There should be no problem here, but these important infections must be thought of and the organisms sought, if early correct diagnoses are to be made.

In general, volumes of pus and pieces of tissue are preferable to swabs and the laboratory should be warned before the specimen is collected. The bacteriologist should use fresh media and appropriate reducing agents and should ensure prolonged, continuous incubation for four or more days in an atmosphere containing 10 per eent carbon dioxide. In assessing results he should pay attention to the amounts of anaerobes, absolutely and relatively to the number of aerobes. The use of 'holding' transport or enrichment media is likely to lead to falacious conclusions about the relative importance of different bacterial species present in a particular specimen.

The following clinical cases are representative of many others we have encountered in dealing with the diagnostic bacteriology for a 550-bed acute hospital during the past 2 years.

CASE 1: PARONYCHIA

A female, aged 41 years presented at hospital with paronychia of the right ring finger of 7 days' duration. The lesion was incised and drained, and Penidural was given intramuscularly. Pus from the lesion yielded a heavy growth of a *Bacteroides* species, sensitive to clindamycin, tetracycline and erythromycin, together with scanty colonies of *Staphylococcus epidermidis*.

CASE 2: ISCHIO-RECTAL ABSCESS

A female, aged 36 years was admitted to hospital with anal pain and dysuria of 6 days' duration. There was a hard indurated, very tender area to the right of the anus - an ischio-rectal abscess. The abscess was incised and drained. Uneventful recovery followed; no antibiotics were given. The pus yielded a pure growth of *Bacteroides melaninogenicus*.

CASE 3: PLEUROPULMONARY INFECTION

A male, aged 36 years, was admitted to hospital at 1500 hrs with sore throat, dysphagia and pyrexia (39.2°C). There was a three-day history of sore throat. On examination, the tonsils were haemorrhagic with free pus, there was mild bilateral tender swelling of the neck, and some palpable cervical glands. Provisional diagnoses of streptococcal tonsillitis, or diphtheria, or infectious mononucleosis were considered, and a throat swab and sample of blood sent to the laboratory. The patient was nursed in isolation, and benzylpenicillin commenced. A routine chest X-ray taken on admission was clear (Fig. 9.1). Three hours later there was a left-sided pleuritic pain with tachypnoea (70 per min), and coarse crepitations were heard over the whole of the left lower lobe. The Paul Bunnell test was negative, and the throat swab yielded only normal flora. There was a leucocytosis of 17,000 (81 per cent polymorphonuclear leucocytes).

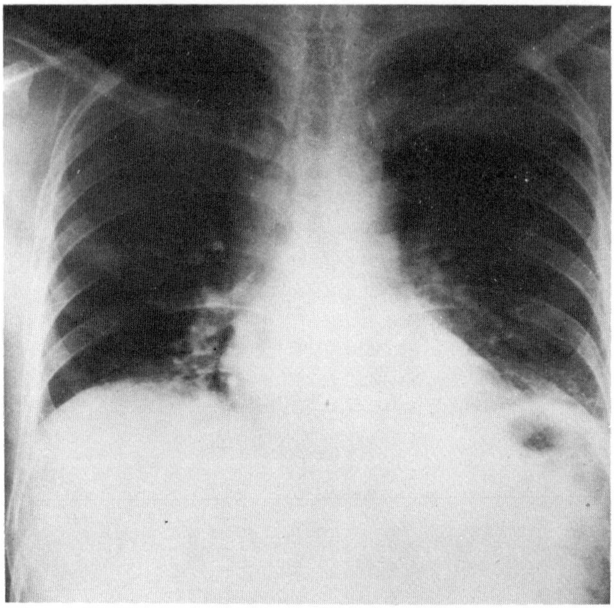

Fig. 9.1. Chest within normal limits on admission (day 1).

At 1300 hrs on the following day, the patient was coughing up a little green sputum, and Ampiclox was commenced. The sputum subsequently yielded a mixed mouth flora with some coliform organisms: anaerobic cultures were not done. He deteriorated rapidly, and by 1700 hrs was very ill, with severe respiratory embarassment. Clinical examination and chest X-ray showed that there was a left-sided hydropneumothorax (Fig. 9.2). Aspiration yielded 600 cm^3 of foul-smelling pus which showed filamentous Gram-negative bacilli and Gram-positive cocci in the direct film. This was regarded as an anaerobic infection, and clindamycin therepy was begun. The patient's condition improved following aspiration.

On the 3rd day, continuous drainage was established at 1300 hrs (Figs 9.3 and 9.4). There was now evidence of

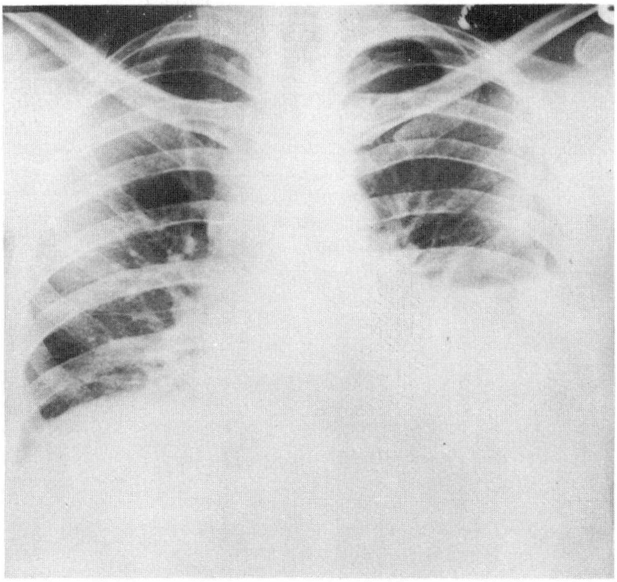

Fig. 9.2. Left sided hydropneumothorax (day 2).

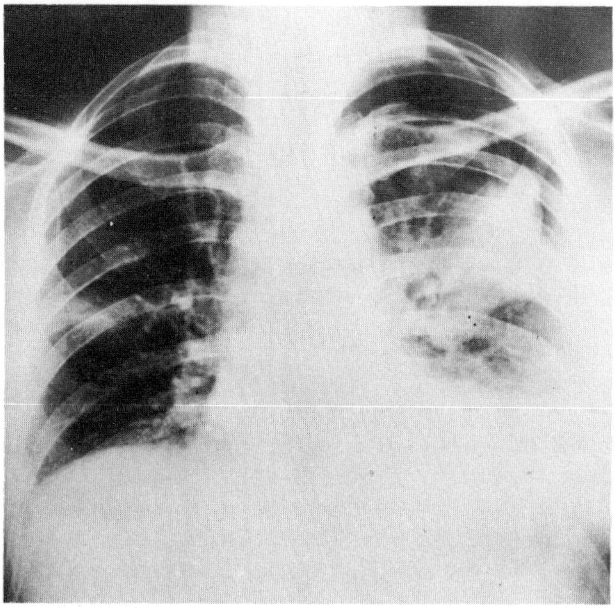

Fig. 9.3. Left-sided pleural thickening and pneumothorax, with bilateral segmental pneumonia (day 3).

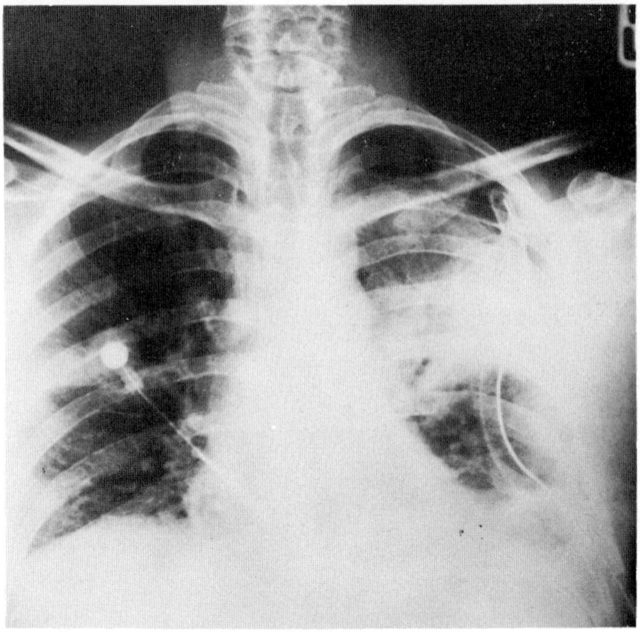

Fig. 9.4. As in figure 3, with drain *in situ* (day 3).

bilateral segmental pneumonia, and the antibiotic therapy was changed to parenteral ampicillin and lincomycin, and oral metronidazole. By 1830 hrs the patient's clinical condition was much improved. At this time, cultures of the empyema pus collected the previous day showed the presence of anaerobes only - a mixed growth of *Bacteroides praeacutus* and anaerobic cocci, sensitive to clindamycin, penicillin, tetracycline, and ampicillin. By the 5th day, the patient was afebrile and much improved, and bacteriological examination of aspirated empyema fluid, which was now much less purulent, showed that the anaerobic flora had been replaced by *Klebsiella aerogenes*.

By the 10th day, the pneumonia had resolved (Fig. 9.5), but there remained an encysted empyema at the left base. Following the institution of adequate drainage, and treatment of the *Klebsiella* infection, there was progressive clearing of the chest, and the patient was finally discharged home on the 36th day.

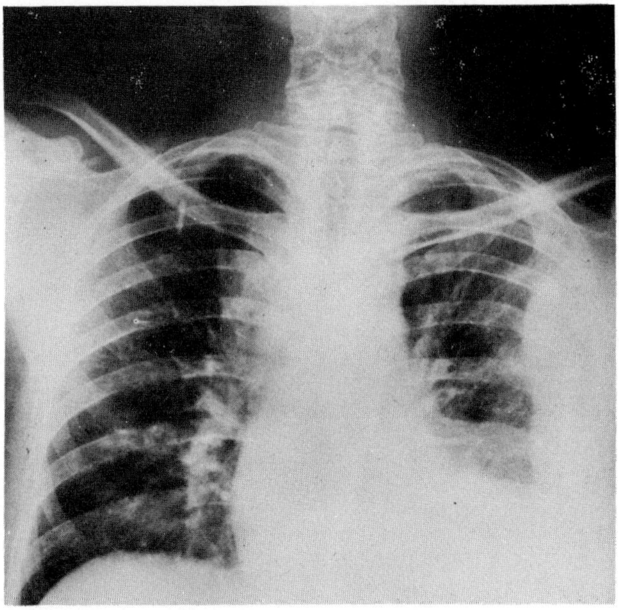

Fig. 9.5. Pneumonia resolved. Encysted empyema at left base (day 10).

CASE 4: POST-APPENDICECTOMY INFECTION.

A female, aged 57 years, was admitted to hospital with a pyrexia of 38°C and clinical evidence of acute appendicitis and pelvic peritonitis. At operation, an acutely inflamed appendix was removed through a right lower paramedian incision. There was a moderate amount of free fluid in the peritoneal cavity, which was drained through a separate drainage tube in the right iliac fossa. Post-operatively, the patient remained pyrexial, and on the 5th post-operative day developed a cellulitis over the left hip and buttock, spreading from the lower end of the laparotomy wound. Ampicillin was commenced on the 7th day, and by the 9th day there was necrosis of the skin and subcutaneous tissues with the formation of multiple subcutaneous abscesses. These were drained surgically, 500 cm^3 of pus expressed from the various incisions, and antibiotic therapy was changed to Ampiclox. Pus sent for bacteriological examination yielded *E. coli;* anaerobic cultures were not done.

Although the patient's general condition remained good, she continued to have a swinging fever, and to discharge foul smelling pus from the drainage sites in the left flank and left iliac fossa. A further abscess which developed on the 16th post-operative day was drained surgically. At this time,

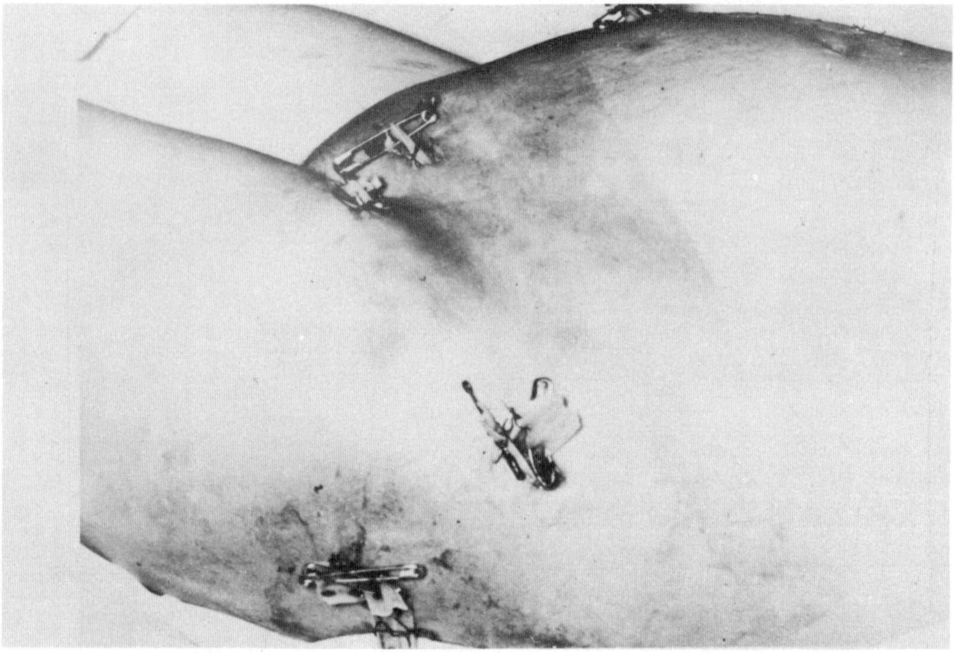

Fig. 9.6. Cellulitis and abscess formation on left side, 16 days after appendicectomy.

an aerobic infection was suspected clinically (Fig. 9.6), and antibiotic therapy was changed to tetracycline. Pus from the various drain sites yielded a predominant growth of *Bacteroides fragilis*, mixed with some *E. coli*. The anaerobe was sensitive to clindamycin, tetracycline, and erythromycin. During the next 5 days the condition subsided. The patient's temperature settled, the incised abscesses dried up, and the cellulitis resolved. Tetracycline was discontinued on the 25th post-operative day, and the patient was discharged to a convalescent hospital.

CASE 5: POST-APPENDICECTOMY PERITONITIS AND SEPTICAEMIA

A female, aged 11 years, was admitted with abdominal pain and vomiting of 12 hrs duration. There was tenderness in the right iliac fossa, and acute appendicitis was diagnosed. At operation the same day a gangrenous appendix was removed. On the 2nd post-operative day the patient vomited after meals and complained of abdominal pain, and there was some rigidity and tenderness about the umbilicus.

There was further deterioration, and on the 4th post-operative day generalized peritonitis was diagnosed. The abdomen was reopened, and copious pus drained from the right iliac fossa and pelvis. Many locules of pus were broken down and sucked out, including a right subphrenic abscess. Peritoneal toilet was carried out, and drains inserted. The patient was put on intravenous ampicillin and transferred to the Intensive Therapy Unit. Pus was sent for bacteriological examination.

The patient remained febrile (40.2°C), the pulse rate rose, and the blood pressure became unrecordable. Gram-negative septicaemia was diagnosed, and intravenous gentamicin was begun. On the 5th post-operative day films from pus collected at operation were seen to contain organisms resembling *Bacteroides*. There was no aerobic growth on direct plates, therefore the presence of an anaerobe was assumed. Gentamicin and ampicillin therapy were discontinued, and intravenous tetracycline commenced.

On the 6th post-operative day the patient's temperature had settled, and she had greatly improved. At this time the presence of a *Bacteroides* species in the abdominal pus was confirmed in culture; the organism was sensitive to clindamycin, tetracycline, and erythromycin.

CASE 6: POST-GASTRECTOMY INFECTION

A female, aged 59 years, with a long standing history of anorexia, exhaustion and abdominal discomfort, was admitted to hospital for blood transfusion (Hb 46%). Five days after admission a high subtotal gastrectomy without splenectomy was performed for gastric ulcer. On the 5th post-operative day there was some offensive discharge from the drain, which, on bacteriological examination, yielded a mixed growth of a faecal streptococcus and *Bacteroides fragilis*, sensitive to clindamycin, tetracycline, and erythromycin.

On the 10th post-operative day the patient was unwell, with fever (38.2°C), sweating and nausea. Blood cultures were taken (these were sterile) and tetracycline therapy commenced. Further deterioration occurred during the next 24 hrs, with swinging fever and the development of left pleuritic pain. A left sided sub-phrenic abscess was diagnosed clinically and radiologically (Fig. 9.7), and this was drained on the 11th post-operative day. About 500 cm^3 of foul smelling pus was removed, and was sent for bacteriological examination. It

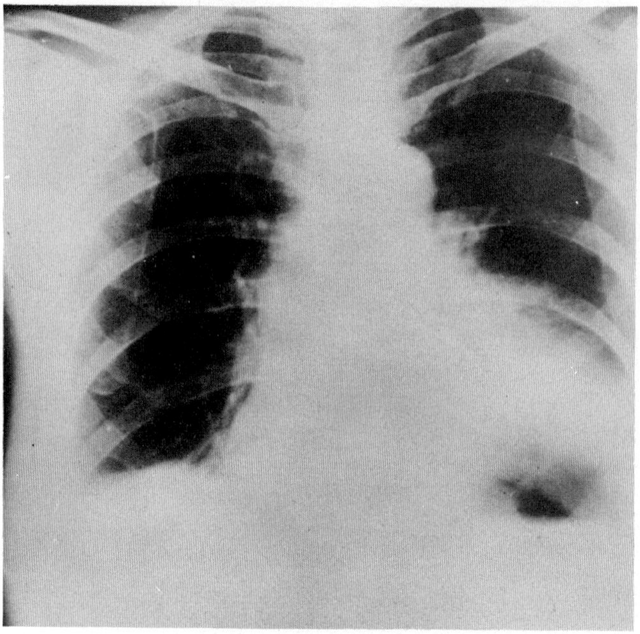

Fig. 9.7. Left subphrenic abscess with fluid level and gas, 11 days after gastrectomy.

later yielded a growth of anaerobes only - predominantly *Bacteroides fragilis*, together with an anaerobic streptococcus. Parenteral tetracycline therapy was commenced immediately post-operatively. There was a rapid improvement in the patient's condition, and oral clindamycin replaced tetracycline on the 13th post-operative day. The temperature returned to normal, the drains dried up, and the patient made an uneventful recovery.

CASE 7: POSTPARTUM INFECTION

Following a normal hospital delivery at term 9 days previously, a woman aged 24 years developed lower abdominal pain and fever (37.8°C). On admission to hospital, there was vaginal bleeding, the cervical os was slightly open, and dark clots were present in the vagina and extruding through the os. There was tenderness generally over the lower abdomen, but the uterus was not palpable through the abdominal wall. A diagnosis of retained products of conception with infection was made; a high vaginal swab was sent for bacteriological examination, and Ampiclox was begun. At evacuation of the uterus there was a large amount of dark clot and minimal retained products. The patient was discharged home apyrexial on the 11th postpartum day. A profuse growth of *Bacteroides fragilis* was obtained from the vaginal swab; it was sensitive to clindamycin, tetracyline and erythromycin.

CASE 8: PARAPUERPERAL INFECTION

Following an uneventful ante-natal period, a primigravida, aged 23 years, was admitted to hospital with ruptured membranes in the 39th week. The patient had been loosing liquor for 3 days before admission, and was pyrexial (37.8°C), but was not in labour. Pelvic examination revealed a breech presentation, cervical of dilation of 1.5 cm and ruptured membrane. A high vaginal swab was sent for bacteriological examination, and the patient was given parenteral ampicillin. With the intravenous infusion of Syntocinon, contractions commenced, and when the cervical os was fully dilated after 4 hours of labour, an assisted breech delivery was performed. The baby was born alive and remained healthy. The patient's temperature settled on the second postnatal day, and there followed an uneventful pureperium. The high vaginal swab yielded a pure growth of *Bacteroides fragilis*, sensitive to clindamycin, ampicillin and erythromycin, but resistant to tetracycline.

CASE 9: TUBO-OVARIAN ABSCESS FOLLOWING STERILIZATION

A married woman, aged 33 years, was referred to hospital with lower abdominal pain, dysuria and frequency, and pyrexia (38.9°C) of one days' duration. Five months previously she had a laparoscopic sterilization performed at another hospital, and for the two weeks before admission she had complained of constipation and dyspareunia. On examination, there was generalized peritonitis, and a tender indurated mass in the right vaginal fornix. The diagnosis was right tubo-ovarian abscess, and intramuscular benzylpenicillin and streptomycin were given.

At laparotomy, the diagnosis was confirmed, pus was evacuated, a right salpingo-oophorectomy performed, and a corrugated drain was inserted through a stab incision in the right iliac fossa. Following operation, the temperature settled rapidly, and the drain was removed on the next day. On the 3rd post-operative day the antibiotic therapy was changed to oral ampicillin, and the patient made an uneventful recovery. Pus from the pelvic abscess yielded a pure growth of *Bacteroides fragilis*, sensitive to clindamycin, ampicillin and erythromycin, but resistant to tetracycline.

COMMENT

In common with others, we have found that many of the infections due to non-sporing anaerobes are associated with operative interference of the gastrointestinal and female genital tracts. Copious pus, often with the formation of large, necrotic, foul-smelling abscesses, is a common denominator of almost all these established infections, for which the treatment of first importance is surgical drainage. It seems likely that in many instances this alone would be sufficient to effect a cure, although it is common practice to prescribe antibiotic therapy also. There is, of course, an absolute and urgent indication for antibiotic therapy in patients who become septicaemic, and antibiotics may also be of value in preventing the development of bacteraemia in patients with established localized infections. In accordance with accepted surgical principles, it is also clear that antibiotics can in no way replace surgical drainage when pus has accumulated in the tissues. Antibiotics might also help to prevent thromboembolic complications that are common in infections with non-sporing anaerobes.

Because many patients with non-clostridial anaerobic infections, and especially those with bacteraemia, are in need of early antibacterial treatment, empirical therapy must begin before a bacteriological diagnosis is made. In our experience, for most non-sporing anaerobic organisms, the most likely effective drugs are clindamycin, lincomycin, tetracycline, and erythromycin, but not penicillin and streptomycin. In order to cover the possibility of an anaerobic infection in a seriously ill patient, and also to guard against the contingency that the infection is due predominantly to a facultative organism, an acceptable empirical regimen is gentamicin combined with clindamycin or tetracycline (but see Chapter 4 about the emergence of tetracycline resistance. Eds.) . The antibiotic therapy may then be modified in the light of the bacteriological findings. The place of metronidazole in the treatment of anaerobic infections other than Vincent's stomatitis has yet to be established, although it seems to be a drug of great promise (Tally *et al.*, 1972).

Acknowledgements

We are greatly indebted to our clinical colleagues who have allowed us to make use of their case notes, and in particular to Mr G.H. Bancroft-Livingstone, Mr G.C. Brentnall, Mr R.V. Fiddian, Mr D.H.C. Harland, Dr B. Harrold, Mr P.R.N. Kerr, Mr R. Rothwell-Jackson and Mr S. Seligman. We thank Dr H.J. Hagger for radiological advice, and Mrs C.L. Bullen, Mrs K. Williams, Mr K. Phillips and Mr P.V. Tearle for technical assistance. We are indebted to Mr J. Harrison and Miss G. Frankland for production of the figures.

REFERENCES

Alston, J.M. (1955). Necrobacillosis in Great Britain. *British Medical Journal*, 2, 1524.

Child, J.A., Darrell, J.H., Davies, N.R. & Davis-Dawson, L. (1969). Mixed infective endocarditis in a heroin addict. *Journal of Medical Microbiology*, 2, 293.

Finegold, S.M. & Rosenblatt, J.E. (1973). Practical aspects of anaerobic sepsis. *Medicine, Baltimore*, 52, 311.

Finegold, S.M., Rosenblatt, J.E., Sutter, V.L. & Attebery, H.R. (1972). *Scope Monograph on Anaerobic Infections*. Upjohn: Michigan.

Gillespie, W.A. & Guy, J. (1956). Bacteroides in intra-abdominal sepsis. *Lancet*, 1, 1039.

Gunn, A.A. (1956). Bacteroides septicaemia. *Journal of the Royal College of Surgeons of Edinburgh*, 2, 41.

Levison, M.E. (1973). The importance of anaerobic bacteria in infectious diseases. *Medical Clinics of North America*, 57, 1015.

Marsden, H.B., & Hyde, W.A. (1971). Isolation of *Bacteroides corrodens* from infections in children. *Journal of Clinical Pathology*, 24, 117.

Mitchell, A.A.B. & Simpson, R.G. (1973). Bacteroides septicaemia. *Current Medical Research and Opinion*, 1, 385.

Okubadojo, O.A., Green, P.J. & Payne, D.J.H. (1973). Bacteroides infection among hospital patients. *British Medical Journal*, 2, 212.

Stokes, E.J. (1958). Anaerobes in routine diagnostic cultures. *Lancet*, 1, 668.

Tally, F.P., Sutter, V.L. & Finegold, S.M. (1972). Metronidazole versus anaerobes. *California Medicine*, 117, 22.

Tracy, O., Gordon, A.M., Moran, F., Love, W.C. & McKenzie, P. (1972). Lincomycins in the treatment of bacteroides infections. *British Medical Journal*, 1, 280.

10. Round Table Discussion

E. JOAN STOKES, A. T. WILLIS, A. PERCIVAL,
J. G. BARTLETT, S. M. FINEGOLD and others

DR STOKES: To begin with I want to make just two little points myself. One is that I am really, as a bacteriologist and not mainly a clinician, against selecting the specimens on which one does anaerobic bacteriology. I think that pus should be examined anaerobically and if this is too expensive for anybody and you can only afford one plate, then use an anaerobic plate. I notice that there is one rather small anaerobic clinical condition which has not so far been mentioned and I suspect it may be because people do not cultivate all specimens anaerobically. That is that one can isolate a pure growth of anaerobic cocci from at least a third of infected sebaceous cysts and although this perhaps is not very devastating to the patients because they are not going to die, it is nevertheless of interest and it is a pity to miss it. The other point I wanted to make is that I think there should be some agreement on what we mean by 'anaerobe'. This is particularly confusing I think in the cocci, both the streptococcal types of anaerobe and the peptococci, because I feel that in some series that are published, some of them are in fact CO_2 dependent cocci and my definition of an anaerobe would be something that does not grow on blood agar within 48 hours plus 5-10% CO_2. I think there is no agreement about this and it would help if people were to state what they mean when they talk about anaerobes.

Dr Alan Percival, who keeps both feet firmly on the ground, will now tell all us clinical bacteriologists what we ought to be doing and what we are missing. I would like to pick up the point that Professor Williams originally raised - the difference between 'isolated from' and 'caused by'. This is the one big void which we really haven't so far answered and I hope that Dr Percival will add a bit in this direction as well.

DR PERCIVAL: I just want to look at the overall situation for someone in a laboratory.

Table 10.1. Percentage of *Bacteroides* isolated from routine specimens, reported in different studies.

Author	Year	No. of +ve cultures	% with *Bacteroides*	Medium* Selective	Enriched
Stokes	1958	4737	5.6	-	-
Smith *et al*	1968	3750	2.6	-	TB
Zabransky	1970	820	26.7	PV	BB
Percival	1974	1125	12.3	N	-

* TB = Thioglycollate broth; BB = Brain broth; PV = Paromomycin-vancomycin; N = neomycin.

Table 10.1 shows you that the people who use selective and enrichment media are the ones who have had the highest overall isolation rates from all the specimens reaching the laboratory. Our Chairman was one of the first, and had 5.6 per cent - these are Gram-negative anaerobes and she used neither selective nor enriched media. The best result was the Zabransky (1970) series from the Mayo Clinic where they do a lot of reparative surgery I understand. They had the highest rate but may have been dealing with a specialised type of material. In our laboratory we find about 12 per cent. Now, most of these patients are not ill and although undoubtedly we have seen some very sick people this is a minority of those from whom we grow anaerobes in profusion. I think our laboratory procedure has got to recover them, because we never have sufficient clinical data on individual specimens, and we must be ready to provide the information when we find out that the patient is or is not ill.

Moving on to sensitivities, as we all know it is better to do agar dilutions or MICs, but by and large a routine laboratory has got to do a disk test. With tetracycline, this shows a sort of regression line that might even satisfy Professor Garrod. If the zone size is more than 25 mm with a 25 µg disk it is sensitive in terms of therapeutically attainable levels of the agent and if it is less, it is resistant. Some strains are more resistant, we find, than Dr Phillips showed. But with something like ampicillin (25 µg disks) the regression lines would not satisfy Professor Garrod and they don't even satisfy me. This is very difficult and it is because many are penicillinase-producers, particularly the resistant or relatively resistant ones.

We have to remember all the time, as our speakers have said repeatedly this afternoon, that these anaerobes are isolated from mixed cultures. There are other organisms there - therefore we have to decide not only whether to treat but, if we do treat, for which of this mixture of organisms do we do so? If we treat for all of them, then I don't think it has been sufficiently brought out that they have a different antibacterial susceptibility pattern from the aerobic enterobacteria with which they are associated.

Now on a point of pathogenicity, and it is related to sensitivity testing, we really don't know whether some of these agents like cotrimoxazole or ampicillin work in the human patient. In the experimental animal, the *Bacteroides* are extremely non-pathogenic, which has not been mentioned so far. You can pump very large numbers of them into a guinea pig, a rabbit's ear, a mouse's peritoneum; they do not touch them on their own. If you combine them with some other agents, sometimes there is some evidence of a symbiotic effect in an experimental model, but the lack of experimental models means that we cannot say 'well at least in the experimental animal on this or that dose, of this or that agent, it works'. There isn't any experimental system to test them in, and this is a great difficulty.

Moving on from pathogenicity, it is quite clear from what has been said that *Bacteroides fragilis*, of the Gram-negative ones, is most commonly isolated even by people who do know how to grow them. We have done some serotyping and the next odd thing about these organisms is that they are poorly antigenic. You can put a very large number of the organisms into rabbits and you only get maximal titres of about 1:1280 but one of my colleagues Mr Rowlands, has managed to produce some fairly specific antisera, and we find that about 70 per cent of isolates in this country were typable with 8 antisera. When we tested some organisms, kindly supplied by Professor R.E.O. Williams from the Far East, only about 18 per cent were agglutinable by the sera, which is not surprising. They have different strains. We also found that the serotypes of organisms from lesions were the same as the preponderant organism in the patient's faeces at the same time. It is a similar situation to that which has been shown, for example, with *Escherichia coli*. But further than that, I think no one has paid tribute to the very great amount of work which has been done by the Germans and Professor Werner for example. According to Werner (1973) *Bacteroides fragilis* and *Bacteroides thetaiotaomicron* are not found in the greatest proportions in the faeces, but they are the only two found from lesions, so his works suggest that of the different, biochemically defined species, *Bacteroides fragilis* is more pathogenic. We don't know yet from our serotyping whether it is particular serotypes that are more pathogenic, certainly some are more common. What has got to be done - because we know that some sick people need treating and we know that the organisms are commonly isolated - is to try and tie these things together in individual cases. We have also to bear in mind, in recommending treatment, that we may have to use a mixture of antibiotics and treat for all, or as some people have suggested, perhaps you need only treat for the anaerobes and ignore the others. These I think are some of the problems that need to be resolved.

DR STOKES: Dr Finegold, would you like to comment at all on the serological reactions in the patient? Has any work been done in the States on this point?

DR FINEGOLD: Not on serology itself, but at the Centre for Disease Control in Atlanta, they have flourescent antibody reagents which will identify the majority of clinical isolates of *B. fragilis* as well as a number of other non-spore forming anaerobes. I think this is one promising approach in terms of rapid bacteriological diagnosis of certain specific organisms.

I might also comment on Dr Percival's discussion of Dr Werner's work. I do think that *Bacteroides fragilis* sub-species *fragilis* and sub-species *thetaiotaomicron* are more commonly involved in clinical infection than the other three recognised sub-species of *B. fragilis*. One problem, however, is that sub-speciation may be difficult. It is based on a few simple fermentation reactions and indole production - but it may be remarkably difficult to reproduce the tests. However, we find that if you go about it carefully and choose certain rigid criteria, then they are reproducible. Indeed, it is those two particular sub-species, *fragilis* and *thetaiotaomicron* which account for the vast majority of clinical infections. One of our Fellows, Dr Casciato, is doing some interesting work which may help to explain why this should be so. He finds that in general *B. fragilis* may be quite susceptible to the killing effect of normal serum in presence of complement but the sub-species *fragilis* and *thetaiotaomicron* are quite resistant to serum bactericidal activity.

DR STOKES: We must now give you all a chance to ask questions.

DR NAKHLA (ST ALBANS): Professor May preaches that he can correlate the pathogenicity of bacteria isolated from the sputum, with the presence of antibodies in the patients' serum. Has Dr Percival looked into this point - the antibody response of patients with *Bacteroides* infection as evidence of pathogenicity in contradistinction to their mere presence as commensals or contaminants?

DR PERCIVAL: We have tried a few patients with what we thought were *Bacteroides* infections - none of them was really critically ill - and we found they had titres which were low by comparison, for example, with enterobacteria. We didn't find agglutinating titres or haemagglutinating titres above 1:80 or 1:160 in these patients and from our findings with animals this didn't surprise us. I cannot yet answer your question whether we can use it as a differential.

DR DAVIES (EDGWARE): Could I ask the panelists how far the ordinary district hospital laboratory should go with its identification of these anaerobes? Our speakers today have been roughly divisible into two - those who have used terms such as 'the anaerobic Gram-negative bacilli and the anaerobic Gram-positive cocci' and those who have used pure speciated terms. How far should we go?

DR PERCIVAL: If you are not doing any studies and you are just trying to give guidance on whether to treat a patient or not, I would say not very far at all. Unless it be shown by someone to an extent that I am not aware of as yet, that the finding of one particular type had a clinical association or connotation for a particular type of infection and was more likely to be significant.

DR STOKES: I would agree with Dr Percival - were you going to disagree?

DR FINEGOLD: No, I wasn't going to disagree very much, but I think that those of you who are involved in any kind of investigation really ought to go all the way and define these organisms as best you can. I think this is very important so that we can eventually get a clear picture as to which specific organisms are involved in which infections and for the clinician to know the general nature of the organism and perhaps its susceptibility. The susceptibility does come into the picture: there are certain anaerobes that are relatively resistant - *Bacteroides fragilis* is the most well known of these but also resistant are *Fusobacterium varium* and *Clostridium ramosum*. So if you know specifically that you are dealing with one of those three species, or if you have susceptibility patterns, it is helpful. I think we will find as time goes on that there are certain organisms of special importance, and therefore at least those workers involved in investigation must go at it wholeheartedly. There is some information out of the CDC group which suggests that *Clostridium septicum* infections tend to be associated with malignancy and we have seen this ourselves in human infections; while this may not be completely documented, I think there is enough to it that if I isolate *C. septicum*, I do go to some pains to exclude an underlying malignancy.

DR STOKES: Speaking as someone who is in rather the same position as Dr Davies, I sympathise with not going the whole hog, although I agree entirely with Dr Finegold that theoretically one should. Ought we not to link up the laboratories who do the primary isolation and give a report within reasonable time that you have isolated a non-sporing anaerobe, and then send that strain to someone who has the expertise and the time to sort it our properly? It is the job of the clinical bacteriologist to link it with the clinical condition of the patient so that the whole thing can really be properly investigated but I don't really see a large number of diagnostic laboratories being able to do the sort of identification which is necessary for these organisms at the moment.

DR FINEGOLD: Why should we look only at anaerobes in this way? Why don't we take the same attitude to facultative and aerobic bacteria - why speciate these? Why isn't it enough to say that this is a Gram-negative rod that is sensitive to this or that? I think we would all be unhappy with that position. It is just the fact that we are pretty accomplished at differentiating these organisms that permits us to go ahead and do it. I grant that the other overriding factor is that it takes more time and more expertise than we have at the moment, but I would like to see us get there eventually. We have on the horizon a number of simplified approaches that will make it feasible for even small clinical laboratories to do a fairly good job of identification. For example, the API system which has been promoted for *Enterobacteriaceae* appears to be applicable to differentiation of various anaerobic organisms as well. The CDC group has done a study on this and several groups, including our own, are studying it at the present time and it really looks quite good. It is inexpensive, easy to handle and, I think, very useful.

DR NEWSOM, (PAPWORTH HOSPITAL, CAMBRIDGE): How often do the members of the panel think when they grow anaerobes, particularly *B. fragilis*, in pure culture that this is a left-over from a mixed lesion which has been inappropriately treated?

DR BARTLETT: *Bacteroides fragilis* is infrequently encountered alone in pus although it is the most common anaerobe in blood cultures. When *B. fragilis* is recovered from pus in pure culture the question may be raised as to whether there have been other more fastidious organisms there also. Another possibility is that an antibiotic had been given which was not active against *Bacteroides fragilis*, and therefore selected the organism from a mixed flora. The correlation with the Gram stain is very important and if this shows no other organisms this would corroborate the culture result. On the other hand, if other bacteria are seen and not grown, there was probably an error in the bacteriological methods.

DR QUAYLE (CARDIFF): May I return to an earlier question, that of prophylactic antibiotic cover for patients with valvular heart disease? Do you cover all patients who have had rheumatic fever - if so with what, and if not, what are the criteria for covering the particular group of patients they do select?

DR BARTLETT: I take it that these are rheumatic fever patients who are undergoing dental procedure? Our recommendation is that of the American Heart Association which is to give penicillin shortly before the procedure and for one or two days after the procedure. The choice of agents is based on the likely organisms that would be involved in a bacteraemia during the procedure and, being oral bacteria, these are usually susceptible to penicillin.

DR. STOKES: And for somebody who is hypersensitive to penicillin?

DR BARTLETT: I would think that in that situation erythromycin would be an adequate substitute, or possibly a cephalosporin.

DR QUAYLE: May I ask if you treat every patient who has had rheumatic fever or only those with clinical evidence of heart disease?

DR BARTLETT: The only ones that we would protect are those with valvular disease.

DR MADDOCKS (LONDON): I should like to raise the subject of pseudomembranous enterocolitis since we have a wide variety of anaerobic expertise present. The prospect of giving anaerobic prophylaxis seems to come up more often day by day and, equally, each week the Lancet brings another crop of letters about this problem. I don't know if anyone has any personal experience or if we could hear about it from our speakers.

DR FINEGOLD: I don't have any personal experience with it but I guess that this is of some interest in its own right because we have used quite a bit of both lincomycin and clindamycin. With lincomycin we saw a lot of diarrhoea - 15 per cent of our patients had diarrhoea and at times this was fairly impressive, although rarely impressive enough to cause us to switch therapy. We have not seen that with clindamycin, but of course I am aware of the report in The Lancet and the letters subsequent to that, and the report in the JAMA with at least one letter subsequent to that. I am also aware of a number of verbal reports, particularly those attributed to Jay Sanford from Dallas, who has had some 20 cases referred to him because of this complication. I don't know what the actual incidence is but the fact that one person could see 20 cases suggests that it is common enough, at times, at least, that we ought to be aware of it and on the lookout. I have no idea what the mechanism is for this disease - I am not aware of any good information which would suggest that this is a specific infection or a specific alteration in normal flora. Presumably, it would be one of those two things rather than a chemical process.

DR PERCIVAL: I could refer you to papers (Reiner, Schlesinger & Miller, 1952; Foster & Dadey, 1959) which show quite clearly that this condition follows treatment with a variety of different antibiotics or none at all. Personally I think it is quite remarkable that someone should publish a paper in a leading journal without apparently making any effort to try and establish the aetiology of the condition. You would have thought, having seen the first two or three cases, when the next ones came along they would at least have made some attempt to do some bacteriological studies. I think it is very disturbing that this is something we should all know about if it is true and yet we are left hanging in the air. None of the correspondence has reported any effort to find any aetiological or microbiological cause either, but I would like to put it in perspective and make the point again that it uncommonly follows treatment with many different wide spectrum antibiotics.

DR SAY (LONDON): I wanted to ask Dr Stokes the methods that she might be using for doing antibiotic sensitivity testing with anaerobes in a routine bacteriological laboratory, which is a problem that we face quite often especially having to wait for some anaerobic Gram-positive rods which sometimes may take four days to grow. And secondly, I would like to ask about the main problem that one faces in the situation of mixed infections - aerobic and anaerobic - in conditions with severe pyrexia, say in gynaecological patients. Do the members of the panel think that trimethoprim-sulphamethoxazole combinations, which usually suit the sensitivity for both organisms, should not be used at this point?

DR STOKES: We use the control disk method for anaerobes as well as for aerobes. The snag that we find with it is that if you get a very slow growing one, you very often wipe the growth off the plate altogether if you try and put four disks on a plate. I frequently have people say that an anaerobe has died but if you test it against just one drug, usually penicillin, and put one disk on the plate, you find, when it finally grows, an enormous zone. It is fair under those circumstances to call it penicillin sensitive but of course if I am really in trouble I am fortunate in having Miss Waterworth who will do a MIC for me. The other question was about the mixture of anaerobes and aerobes in gynaecological infections. Who would like to cope with that?

DR WILLIS: This depends how ill the patient is and whether the condition is going to be amenable to surgery. There is no doubt that in patients, say with a tubo-ovarian abscess, a pelvic abscess of any sort, or intrauterine infection, if surgery is conducted, this often leads to a very dramatic resolution of the patient's fever. For those patients in whom one suspects that there is a Gram-negative aerobic element as well as an anaerobic element and the patient is extremely ill, perhaps bacteraemic, then our procedure, rightly or wrongly, is to give gentamicin and clindamycin, or lincomycin, as soon as we have got our bacteriological specimens. Twenty-four hours later we are in a position to judge, tentatively, what the position is in terms of the aerobic culture. If we have nothing very significant or very minimal growth on the aerobic culture, we still assume it is an anaerobic infection, but we may have to wait for another 24 hours to be sure. If we get nothing on the aerobic cultures we assume that it is definitely anaerobic and we drop out the gentamicin at that point.

PROFESSOR BRUMFITT (ROYAL FREE HOSPITAL): I would like to return to the question of enterocolitis - I think it is important that physicians report the occurrence of this condition if it is seen with undue frequency after the administration of an antibiotic, especially if it has recently been introduced into clinical practice. I would assume that the authors of the various reports on this topic had already excluded the presence of common pathogens in the faeces (such as staphylococci). For example, if an experienced clinician like J.P. Sanford has information on 20 cases, surely investigations of an elaborate kind would have been done by him.

DR FINEGOLD: I know he looked specifically for the *Staphylococcus* and didn't find it.

PROFESSOR BRUMFITT: In view of the remarkable number of cases of enterocolitis that have occurred following the use of clindamycin, and the known tendency of lincomycin to cause diarrhoea, these new reports relating it to clindamycin must be investigated.

The association of tetracycline and staphylococcal enterocolitis is well documented but on the other hand one must remember that literally tons of sulphonamides have been given to patients without causing enterocolitis.

The next question I would like to ask is why there appears to be an objection to using rifampicin, rather than clindamycin, in conjunction with gentamicin, for the treatment of Gram-negative septicaemia, especially if the infection is believed to originate from the gall bladder.

Also if the anaerobic infection is associated with staphylococcal infection and staphylococcal septicaemia, the evidence that clindamycin is not a very efficient agent against the staphylococcus when the bacterial population is high must be taken into account (in other words with increasing numbers of staphylococci the efficiency of clindamycin may be reduced). Would the panel please comment?

DR STOKES: Which of you would like to say something about that?

DR FINEGOLD: I might just comment on the question of rifampicin plus gentamicin. The possible problem is that in any given population of anaerobes, or at least in several that we have looked at, there seems to be a number of resistant cells, so that if one used rifampicin as a single agent, these might be selected out and take over. In effect, when using it with gentamicin, as far as the anaerobes are concerned, you would be using it as a single agent. Further than that, there is no clinical experience with it although I have no doubt that it would be good, based on *in vitro* data, and its use in other diseases.

DR STOKES: Thank you very much Dr Finegold. I am afraid we must stop the questions now. I would like those of you who tend to be too enthusiastic about anaerobes - and I am sure we are all, after today's symposium, extremely enthusiastic about them - to read The New England Journal of Medicine which says 'Beware the Anaerobic Bandwagon' and I'll leave you with that thought. I must thank all the speakers very much for their excellent contributions and I would also like to thank Upjohn for their most generous support of this meeting.

REFERENCES

Foster, F.P. & Dadey, J.L. (1959). Pseudomembranous Enterocolitis, a growing menace. *Antibiotics Annual 1958-59*, 757.

Reiner, L., Schlesinger, M.J. & Miller, G.M. (1952). Pseudomembranous colitis following Aureomycin and Chloramphenicol. *Archives of Pathology*, 54, 39.

Smith, C.D., Deane, C., Montgomery, J., Williams, R. & Sampson, C.C. (1968). Bacteroides; incidence and studies on 100 isolations. *Journal of the National Medical Association*, 60, 215.
Stokes, E.J. (1958). Anaerobes in routine diagnostic cultures. *Lancet*, 1, 668.
Zabransky, R.J. (1970). Isolation of anaerobic bacteria from clinical specimens. *Mayo Clinic Proceedings*, 45, 256.

11. Appendices

APPENDIX A

CULTURE MEDIA AND REAGENTS

In this appendix we give the commercial sources for the U.K. and in some cases details of composition, of less familiar culture media and reagents recommended by contributors to the symposium. Further information will be found in the relevant chapters of this book, and in the publications listed in Appendix D.

1. Transport Media

Cary and Blair Transport Medium (recommended in U.S.A.)	Turner-Stayne Laboratories Ltd., (Becton Dickinson) York House, Empire Way, Wembley, Middlesex.

2. Isolation Media

Dr B. Drasar. There is no universally satisfactory medium for the growth of non-sporing anaerobes. A useful basal medium consists of: 1% peptone, 1% Yeast extract, 0.4% Mineral salt solution, 0.0005% Hemin, 0.05% L-cysteine + 0.03% sodium formaldehyd sulphoxalate.

The mineral salt solution has the following composition: $CaCl_2$ (anhydrous) 0.2g; $MgSO_4$ 0.2g; K_2HPO_4 1.0g; KH_2PO_4 1.0g; $NaHCO_3$ 10.0g; NcCl 2.0g. Mix $CaCl_2$ and $MgSO_4$ in 300 ml distilled water until dissolved. Add 500 ml water and, while swirling, slowly add remaining salts. Continue swirling until all these salts are dissolved. Add 200 ml distilled water, mix and store at 4°C.

This medium may need to be fortified with rumen fluid (available from Tissue Culture Services, Slouth) or 10 per cent ox bile (Oxoid Ltd) for growth of some strains of *Bacteroides*. Tween 80 is stimulatory for some propionibacteria and anaerobic cocci.

Reinforced Clostridial Agar with 10 per cent horse blood or egg yolk is also a useful simple isolation medium. After storage for 3 days in an anaerobic cabinet, commercial ready-poured plates can be used.

See also Watt and Collee (Chapter 2), and, for selective media, Phillips (Chapter 4).

3. Biochemical Tests

Thioglycollate Fermentation Base	Turner-Stayne Laboratories Ltd (Becton Dickinson) York House, Empire Way, Wembley, Middlesex.
API system for anaerobes	API Products Ltd., Philpot House, Rayleigh, Essex.
4. *Antibiotic sensitivity discs* for identification (see Ch. 4)	Diamed Diagnostics, 24-36 Queensland Street, Liverpool, L7 3JQ.

APPENDIX B

Apparatus

1. Anaerobe Jars

Standard	Baird and Tatlock (London) Ltd., P.O. Box 1, Freshwater Road, Romford R.M.1. 1HA.
Large	Turner-Stayne Laboratories Ltd., (Becton Dickinson) York House, Empire Way, Wembley, Middlesex.
Milking machine pails (for conversion)	Fullwood and Bland Ltd., 35 Bevenden Street, London N.1.
2. Anaerobe Cabinets	Plysu Industrial Ltd., Woburn Sands, Bletchley, Bucks.
	Vickers Limited, Pristley Road, Basingstoke, Hampshire.
Rigid plastic glove boxes and air-locks suitable for flexible plastic isolators	Lintott Engineering Ltd., Foundry Lane, Horsham, Surrey.
	Henry Hartley & Co, Ltd., London W.12.

Airlocks for Anaerobic Cabinets

The airlock of an anaerobic cabinet must be able to withstand the stress resulting from its partial evacuation. It is difficult to obtain suitable airlocks of sufficient size to allow the passage of an anaerobic jar. The advice of the manufacturer should be saught with respect to any particular product.

The large airlock illustrated in Fig. 3.1 was manufactured by Henry Hartley and Co. Ltd., London W.12. This airlock which has an internal diameter of 1 ft 6 inches and overall length of 3 feet is constructed from stainless steel plate, and the doors are 1 inch thick perspex.

A similar airlock could be constructed by many engineering firms. Alternatively an aluminium pressure barrel could be converted. Guiness barrels have an internal pressure chamber and are difficult to convert.

3.	V.P.I. Anaerobic Culture System: Gassing stand, clamps, Roll tube spinner	Bellco Glass, Inc., 340 Edrudo Road, Vineland, New Jersey, 08360, U.S.A.
4.	Oxygen-Free Gas for P.R.A.S. Culture System	B.O.C. Special Gases Division, Deer Park Road, London, S.W.19.
		Air Products Special Products Dept., Weston Road, Crewe, Cheshire, CW1 1DF.
5.	Butyl rubber tubing and bungs (PRAS system)	Esco Rubber Co. 14, Great Portland Street, London W1.
6.	Tissue-culture rack with toggle-clipped lid	Denley Instruments Ltd. Bolney Cross, Bolney, Sussex.
7.	GasPak system	Turner-Stayne Laboratories Ltd., (Becton Dickinson) York House, Empire Way, Wembley, Middlesex.
8.	Gas Chromatography	Pye Unicam Ltd., York Street, Cambridge, CB1 2PX.
9.	Microtiter	Flow Laboratories, Victoria Park, Heather house Road, Irvine.
	Autotiter	Fisons Scientific Apparatus, Bishop Meadow Road, Loughborough.

APPENDIX C

KEYS TO THE GENERA OF ANAEROBIC BACTERIA AND SOME IMPORTANT BIOCHEMICAL REACTIONS OF NON-SPORING ANAEROBES ISOLATED FROM HUMAN CLINICAL SOURCES.

Abstracted from: (Holdeman & Moore 1972), Anaerobe Laboratory Manual, with the kind permission of Dr W.E.C. Moore.

Gas Chromatography. Space does not permit a detailed discussion of results of gas chromatography. Those interested are referred to the Anaerobe Laboratory Manuals of Holdeman & Moore, and Sutter *et al.* (see Appendix D).

Legend of symbols

+ = positive reaction for 90-100% of strains
- = negative reaction for 90-100% strains
a = strong acid - pH 5.5 or below - for 90-100% of strains*
c = curd (milk)
d = digested (milk, meat, casein)
s or S = stimulated (growth)
v = variable (strains may be either + or -)
w = weak acid - pH 5.5 to 6.0 - for 90-100% of strains*
+, 2, 3, 4 increasing amounts of growth on bile media
$-^{w}$ = most strains negative, some strains weak
$-^{a}$ = most strains negative, some strains acid
a^{w} = most strains acid, some strains weak
$+^{-}$ = most strains positive, some strains negative
w^{a} = most strains weak, some strains acid
$-^{+}$ = most strains negative, some strains positive
a^{-} = most strains acid, some strains negative
w^{-} = most strains weak, some strains negative
I = normal flora, intestine
U = normal flora, urogenital tract
M = normal flora, mouth
S = normal flora, skin

Note aesculin results for ferirc ammonium citrate added after incubation to liquid aesculin medium.

* Uninoculated xylose and arabinose, under CO_2, are often pH 5.9. For cultures in these media, consider pH 5.4 to 5.7 weak acid and pH below 5.4 strong acid.

Table 11.1. Key to Genera of Anaerobic Cocci

COCCI

Gram +ve		Gram -ve	
Requirement for fermentable carbohydrate	No requirement for fermentable carbohydrate	Propionic acid plus acetic acid produced	Butyric acid plus acetic acid produced
Ruminococcus	*Peptostreptococcus* *Peptococcus*	*Veillonella*	*Acidamorococcus*
	(Table 11.2)	(Table 11.2)	(Table 11.2)

Table 11.2

Anaerobic cocci

	Veillonella parvula	Veillonella alcalescens	Sarcina ventriculi	Peptostreptococcus productus	Ps. parvulus	Ps. micros	Ps. intermedius	Ps. anaerobius	Peptococcus variabilis	P. saccharolyticus	P. prevotii	P. morbillorum	P. magnus	P. constellatus	P. asaccharolyticus	Megasphaera elsdenii	'Gaffkya' anaerobia	Acidaminococcus fermentans
cellobiose	-	-	$-^{a}$	a	-	-	a	-	-	-	-	$-^{w}$	-	a^{w}	-	-	$-^{w}$	-
fructose	-	-	a	a	-	-	a	w^{-}	$-^{a}$	a	-	$-^{a}$	-	a	-	a^{-}	a^{-}	-
glucose	-	-	a	a	a	$-^{w}$	a	w^{a-}	$-^{a}$	a	$-^{w}$	w^{a}	$-^{w}$	a	$-^{w}$	a	a	$-^{w}$
lactose	-	-	a	a	a	-	a	-	-	-	-	-	-	-	-	-	-	-
maltose	-	-	a^{w}	a	-	-	a	$-^{a}$	-	-	-	a^{w}	-	a	-	a^{-}	a^{w}	-
sucrose	-	-	a	a	-	-	a	-	-	-	-	w	-	a	-	$-^{a}$	a^{w}	-
gelatin	$-^{w}$	$-^{w}$	-	-	w	-	$-^{w}$	$-^{w}$	+	-	$-^{+}$	-	w^{-}	-	$-^{w}$	$-^{w}$	$-^{w}$	$-^{w}$
indole	-	-	-	-	-	-	-	-	-	-	-	-	-	-	+	-	-	-
nitrate	+	+	$+^{-}$	-	-	-	$-^{+}$	$-^{+}$	-	+	$-^{+}$	-	-	v	-	-	$-^{+}$	-
catalase	-	+	-	-	-	$-^{w}$	-	-	-	+	$-^{+}$	-	-	-	-	-	-	-
human-clinical	+	+	.	+	+	+	+	+	+	.	+	+	+	+	+	.	+	+
-normal flora	IUM	IUM	I	I	.	.	IU	UM	U	S	U	.	.	U	IU	U	U	IU

Table 11.3. Key to the Genera of Anaerobic Gram-Negative Bacilli.

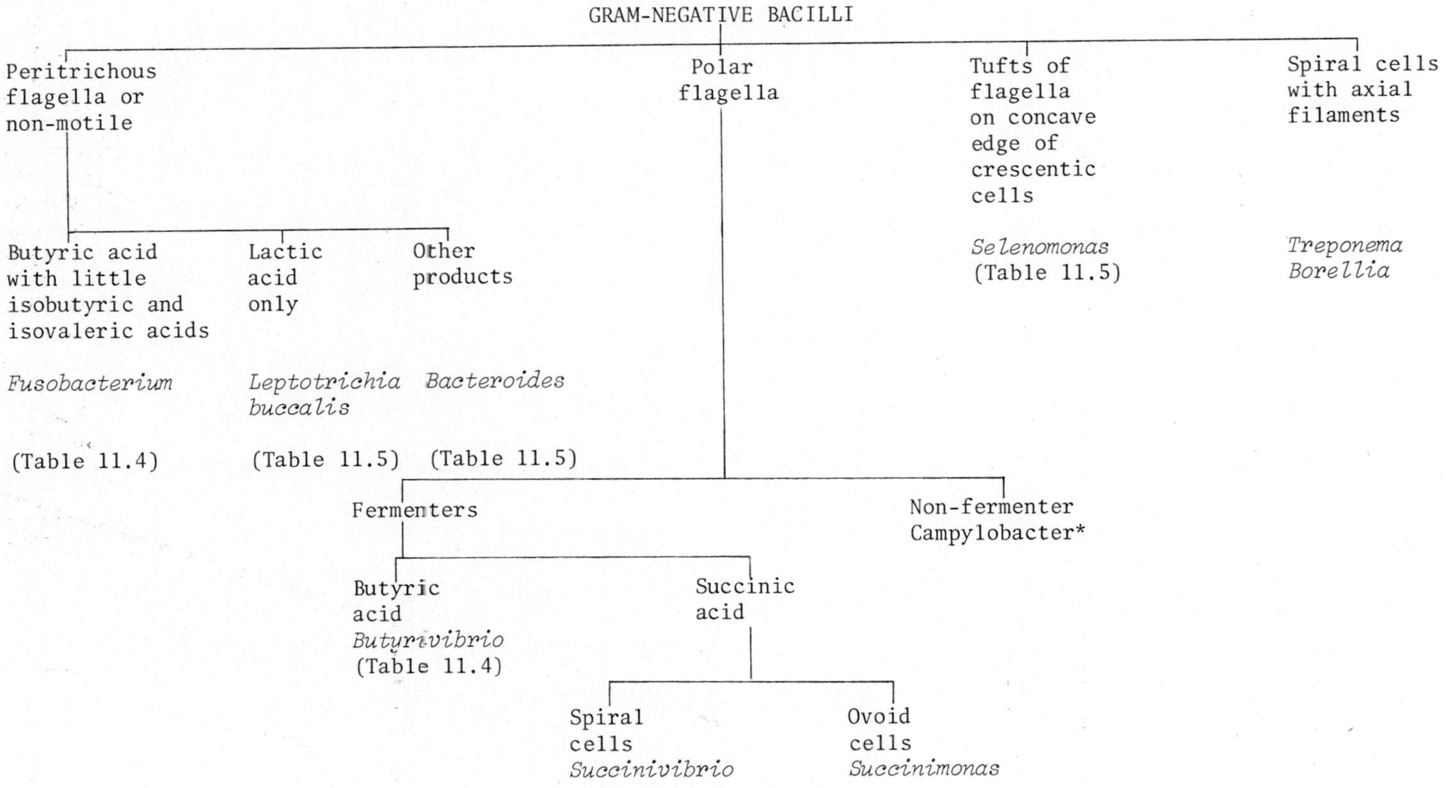

* Requires 10% oxygen

Table 11.4

Fusobacterium, Butyrovibrio

	Butyrivibrio fibrisolvens	F. varium	F. symbiosum	F. russii	F. prausnitzii	F. plauti	F. nucleatum	F. aquatile	F. necrophorum	F. necrogenes	F. naviforme	F. mortiferum	F. gonidiaformans	F. glutinosum	F. bullosum
amygdalin	v	-	-	-	$-^{w}$	-	-	w	-	-	-	$-^{w}$	-	-	-
arabinose	a^{w-}	-	$-^{a}$	-	-	w^{-}	-	w^{a}	-	$-^{w}$	-	-	-	-	-
cellobiose	a^{w}	-	$-^{a}$	-	$-^{w}$	$-^{w}$	-	a	-	w^{-}	-	w^{-}	-	-	-
aesculin hyd.	+	-	-	-	+	-	-	+	-	+	-	+	-	-	-
fructose	a^{w}	w	a^{w}	-	w^{-}	$-^{w}$	w^{-}	a	$-^{w}$	w^{a}	-	w^{a}	-	a	-
glucose	a^{w}	w^{a}	a^{w}	$-^{w}$	w^{-}	w^{a}	$-^{w}$	a^{w}	$-^{w}$	a^{w}	$-^{a}$	a^{w}	w^{-}	a	-
lactose	a^{-}	-	$-^{w}$	-	w^{-}	-	-	a^{w}	-	-	-	w^{a}	-	a	-
maltose	a^{w}	-	-	-	w^{-}	-	-	a^{w}	-	-	-	w^{-}	-	w	-
mannitol	-	-	w^{-}	-	-	-	-	-	-	-	-	-	-	-	-
mannose	a^{w}	w	a^{-}	-	-	$-^{w}$	-	a	-	w^{a}	-	w^{a}	-	-	-
melezitose	-	-	-	-	-	w^{-}	-	-	-	-	-	-	-	-	-
raffinose	$-^{a}$	-	-	-	$-^{w}$	-	-	w	-	$-^{w}$	-	v	-	-	-
rhamnose	v	-	$-^{w}$	-	$-^{w}$	-	-	-	-	-	-	-	-	-	-
starch pH	w^{a}	-	-	-	$-^{w}$	a	-	w	-	-	-	$-^{w}$	-	-	w
sucrose	a^{-}	-	-	-	$-^{w}$	-	-	a	-	w^{-}	-	v	-	a	-
trehalose	w^{-}	-	$-^{w}$	-	$-^{w}$	-	-	-	-	w^{-}	-	$-^{w}$	-	-	-
xylose	a^{-}	-	-	-	-	-	-	w^{a}	-	-	-	$-^{a}$	-	-	-
gelatin	$-^{d}$	$-^{w}$	$-^{w}$	$-^{w}$	-	-	$-^{w}$	-	v	-	-	-	w^{-}	-	-
milk	c^{ad}	-	$-^{c}$	-	-	-	-	c	c^{d-}	-	-	-	-	c	-
indole	-	$+^{-}$	-	-	-	-	+	+	+	-	+	-	+	+	-
nitrate	-	-	-	-	-	+	-	-	-	-	-	-	-	-	-
bile growth	$-^{4}$	4	v	3^{-}	3^{-}	$+^{2}$	$-^{2}$	S	$-^{4}$	4	$-^{2}$	4^{s}	-	4	2
motility	+	-	v	-	-	$+^{-}$	-	+	-	-	-	-	-	-	+
human - clinical	+	+	+	+	.	.	+	.	+	.	+	+	+	+	.
- normal flora	I	I	I	I	I	I	M	I	.	I	M	IU	.	.	I

Table 11.5. *Bacteroides, Leptotrichia, Selenomonas*

	B. biacutus	B. capillosus	B. clostridiiformis ss. clostridiiformis	B. clostridiiformis ss. girans	B. coagulans	B. corrodens	B. fragilis ss. distasonis	B. fragilis ss. fragilis	B. fragilis ss. ovatus	B. fragilis ss. thetaiotaomicron	B. fragilis ss. vulgatus	B. fragilis - other	B. furcosus	B. melaninogenicus ss. asaccharolytucus	B. melaninogenicus ss. intermedius	B. melaninogenicaus ss. melaninogenicus	B. ochraceus	B. oralis	B. pneumosintes	B. praeacutus	B. putredinis	B. ruminicola ss. ruminicola	B. ruminicola ss. brevis	Leptotrichia buccalis	Selenomonas sputigena
amygdalin	-	$-^{w}$	$-^{w}$	$-^{w}$	-	-	w^{a}	w^{a}	w^{a}	a^{w}	v	v	-	-	w^{-}	a^{-}	a	a^{w}	-	-	-	a^{-}	a^{w}	a^{-}	-
arabinose	w^{-}	-	w^{-}	a^{w}	-	-	a^{-}	-	w^{a}	a^{w}	a^{-}	v	-	-	$-^{w}$	w^{-}	$-^{w}$	-	-	-	-	v	a^{w}	$-^{a}$	v
cellobiose	a	$-^{w}$	v	$-^{a}$	-	-	v	$-^{w}$	w^{a}	a^{-}	$-^{w}$	v	-	-	-	a^{-}	w^{a}	a^{-}	-	-	-	v	a^{w}	a	$-^{w}$
aesculin hyd.	+	+	+	+	-	-	+	+	+	+	$+^{-}$	+	+	-	-	$+^{-}$	+	+	-	-	-	+	+	+	$+^{-}$
fructose	a	$-^{w}$	w^{a}	a^{w}	-	-	a^{w}	a^{w}	a^{w}	a^{w}	a^{w}	a^{w}	w	-	v	a	w^{a}	a^{w}	-	-	-	a^{w}	a^{w}	a	w^{-}
glucose	a	a^{-}	a^{w}	a	-	-	a^{w}	a^{w}	a^{w}	a	a^{w}	a^{w}	w^{a}	-	a^{w}	a	a	a^{w}	-	-	$-^{w}$	a^{w}	a	a	a^{w}
lactose	a	$-^{w}$	w^{a}	v	-	-	a^{w}	w^{a}	a^{w}	a^{w}	a^{w}	w^{a}	-	-	$-^{a}$	a	a^{w}	a^{w}	-	-	-	a^{w}	a^{w}	a	v
maltose	w^{a}	$-^{w}$	v	a^{w}	-	-	a^{w}	a^{w}	a^{w}	a^{w}	a^{w}	w^{a}	-	-	v	a	v	a^{-}	-	-	-	$-^{a}$	a^{w}	a	w
mannitol	-	-	-	$-^{w}$	-	-	-	-	w	-	-	$-^{w}$	-	-	-	-	-	-	-	-	-	$-^{w}$	-	$-^{w}$	w^{-}
mannose	a	$-^{w}$	v	a	-	-	a^{w}	a^{w}	a^{w}	a^{w}	a^{w}	w^{a}	-	-	v	a	a^{w}	a^{w}	-	-	$-^{w}$	a^{w}	a	a	a^{w}
melezitose	w^{a}	-	-	v	-	-	v	-	$-^{w}$	v	-	$-^{w}$	-	-	-	-	-	$-^{w}$	-	-	-	-	$-^{w}$	a^{-}	-
raffinose	a	-	v	a^{w}	-	-	a^{w}	a^{w}	a^{-}	a^{w}	a^{w}	v	-	-	$-^{a}$	a	w^{a}	a^{w}	-	-	-	$-^{a}$	a^{w}	$-^{a}$	v
rhamnose	a	-	v	a^{w}	-	-	v	-	v	v	a^{w}	v	-	-	-	$-^{w}$	$-^{w}$	$-^{a}$	-	-	-	$-^{w}$	$-^{a}$	-	w^{-}
starch pH	w^{-}	$-^{w}$	$-^{w}$	w^{-}	-	-	w^{a}	a^{w}	w^{a}	a^{w}	a^{w}	w^{a}	-	-	a^{w}	a	a^{w}	a	-	-	-	v	a^{w}	$-^{w}$	w^{-}
sucrose	a	$-^{w}$	a^{w}	a	-	-	a^{w}	a^{w}	a^{w}	a^{w}	a^{w}	a^{w}	v	-	v	a	a^{w}	a^{w}	-	-	-	a^{-}	a	a	a^{w}
trehalose	w	$-^{w}$	$-^{w}$	w^{-}	-	-	w^{a}	-	v	a^{-}	-	a^{-}	-	-	$-^{w}$	-	-	$-^{w}$	-	-	-	$-^{w}$	$-^{a}$	$-^{a}$	v
xylose	a	$-^{w}$	v	a^{w}	-	-	a^{w}	a^{w}	a^{w}	a^{w}	a^{w}	a^{-}	-	-	$-^{w}$	$-^{w}$	$-^{w}$	$-^{w}$	-	-	-	$-^{a}$	a^{-}	-	a
gelatin	w^{-}	$-^{w}$	$-^{w}$	-	+	-	$-^{+}$	$-^{w}$	$-^{w}$	$-^{w}$	$-^{+}$	$-^{+}$	$-^{w}$	+	+	+	w^{-}	$+^{-}$	$-^{w}$	+	+	$-^{+}$	$+^{w}$	-	$-^{w}$
milk	c	$-^{c}$	$-^{c}$	c	c^{d}	-	c	c	c	c	c	c	-	d^{-}	d^{c-}	c	c	c^{-}	-	-	d^{c}	c^{-}	c^{d}	c^{-}	c
indole	-	-	$-^{+}$	-	+	-	-	-	+	+	-	v	-	$+^{-}$	$+^{-}$	-	-	-	-	-	+	-	$-^{+}$	-	-
nitrate	+	-	v	+	-	+	-	-	-	-	-	$-^{+}$	-	$-^{+}$	-	-	-	-	-	+	-	$-^{+}$	-	-	-
bile gr.	3^{4}	$+^{s}$	4^{-}	4	$+^{s}$	+	S^{4}	S^{4}	S^{4}	3^{s}	S^{4}	4^{s}	$+^{2}$	$-^{+}$	2^{-}	$-^{3}$	2^{-}	$-^{2}$	$-^{+}$	+	3^{-}	-	$-^{2}$	4^{-}	-
motility	-	-	-	+	-	-	-	-	-	-	-	-	-	-	-	-	-	-	-	+	-	-	-	-	+
human																									
- clinical	+	+	+	+	+	+	+	+	.	+	+	+	+	+	+	.	.	+	+	+	+	+	+	.	+
- normal flora	.	I	I	.	I	M	I	I	I	IU	I	IUM	I	I	.	M	M	IUM	.	.	I	.	I	IM	.

Table 11.6. Key to the Genera of Anaerobic Gram-Positive Bacilli

GRAM-POSITIVE BACILLI

- Spores produced: *Clostridium*
- Spores not produced:

Major products propionic and acetic acids	Acetic plus lactic acid	Major product lactic acid	Major product succinic acid*	Butyric acid plus others or no major product
Propionibacterium *Arachnia*	*Bifidobacterium*	*Lactobacillus*	*Actinomyces*	*Eubacterium* *Lachnospira*
(Table 11.7)	(Table 11.8)	(Table 11.9)	(Table 11.7)	(Table 11.10)

* depending on CO_2 concentration

Table 11.7.

Arachnia, Propionibacterium, Actinomyces

	Arachnia propionica	Propionibacterium acnes	P. avidum	P. freudenreichii ss. freudenreichii	P. freudenreichii ss. shermanii	P. granulosum	P. jensenii	P. lymphophilum	Actinomyces bovis	A. israelii	A. naeslundii	A. viscosus
adonitol	-	$-^{a}$	v_{w}	v_{w}	a^{w}	$-_{w}$	a^{w}_{-}	a_{-}	$-_{-}$	$-_{-}$	$-_{w}$	$-_{w}$
amygdalin	v_{-}	$-_{w}$	-	$-_{w}$	$-_{w}$	$-_{w}$	w	w	w_{-}	a_{a}	a_{w}	a
arabinose	w_{a}	$-_{a}$	v_{w}	a_{-}	a_{a}	-	v	-	w_{-}	-	$-_{a}$	$-_{w}$
cellobiose	-	$-_{a}$	$-_{w}$	w	$-_{w}$	-	v_{w}	-	w_{-}	a	w	-
erythritol	-	-	a	v	a	-	a_{w}	a	w_{-}	-	$-_{a}$	-
aesculin pH	-	-	-	-	-	-	-	-	w	v	$-_{-}$	-
aesculin hyd.	-	$-_{-}$	+	+	$+_{w}$	-	+	$-_{-}$	v_{-}	+	$+_{w}$	v_{w}
fructose	a_{w}	a_{w}	a	a	a	a	a	a_{w}	w_{w}	a	a	a
glucose	a_{a}	a	a	a_{-}	a	a	a_{w}	a	a	a	a_{w}	a_{w}
glycogen	-	$-_{w}$	$-_{a}$	w	$-_{-}$	-	-	-	v	-	-	$-_{w}$
inositol	-	-	-	v	a	-	v	v	$-_{w}$	a	v	a
lactose	v	$-_{w}$	v	$-_{-}$	a	$-_{-}$	v_{-}	-	a_{-}	a	a_{w}	a_{w}
maltose	v	-	a_{a}	w	w	a_{a}	a	a	w_{-}	a	a_{a}	a
mannitol	v_{-}	v_{-}	-	$-_{w}$	-	$-_{w}$	a	-	w	a	$-_{w}$	-
mannose	a_{w}	a	a_{-}	a	a_{-}	a	a_{w}	-	w_{-}	a	a	a_{a}
melezitose	$-_{-}$	-	a	-	w	v	a_{w}	-	w_{-}	v_{-}	-	$-_{a}$
melibiose	w	-	v_{a}	-	v	v	a_{w}	-	w	a_{w}	v	-
raffinose	v	-	-	-	-	v	a	-	$-_{-}$	a_{-}	a	a
rhamnose	-	$-_{a}$	$-_{w}$	$-_{w}$	-	-	$-_{w}$	-	w_{-}	w_{-}	$-_{w}$	$-_{a}$
ribose	v_{a}	-	a	a_{w}	v	v_{w}	a_{-}	a	w	a	-	$-_{-}$
salicin	$-_{a}$	$-_{a}$	v_{a}	a	v_{a}	-	a_{a}	-	$-_{-}$	a_{a}	v_{a}	a
sorbitol	-	-	$-_{a}$	-	$-_{-}$	$-_{w}$	-	$-_{-}$	w	-	-	$-_{a}$
starch pH	-	w	-	-	w	-	v	w	a	v	v	w
starch hyd.	$-_{-}$	$-_{a}$	-	$-_{-}$	$-_{a}$	$-_{w}$	$-_{w}$	v	+	-	-	v
sucrose	a	$-_{a}$	a_{w}	w	-	a_{-}	a_{w}	v	v_{-}	a	a_{-}	a
trehalose	a_{w}	$-_{w}$	a_{a}	$-_{w}$	$-_{-}$	w_{w}	a	-	w_{-}	a	a	v_{-}
xylose	-	$-_{-}$	-	-	w_{w}	$-_{-}$	v	-	w	a_{w}	v_{w}	w_{w}
gelatin	v_{d-}	$+_{c-}$	$+_{c-}$	-	$-_{-}$	w_{-}	$-_{c}$	w	-	-	$-_{-}$	-
milk	c	d	d_{d}	-	c	c	-	-	c	c	c	c
meat	-	$-_{-}$	-	-	-	-	-	-	-	-	-	-
indole	-	$+_{-}$	-	-	-	-	-	-	-	$-_{+}$	$-_{-}$	-
nitrate	+	$+_{-}$	-	+	-	-	-	v	-	-	+	+
catalase	-	+	+	w	+	+	v	v	-	-	-	+
human												
- clinical	+	+	+	+	+	+	+	+	.	+	.	+
- normal flora	M	IS	S	.	.	IS	M	.	.	.	.	M

Table 11.8.

Bifidobacterium

	B. adolescentis variety A	B. adolescentis variety B	B. adolescentis variety C	B. adolescentis variety D	B. bifidum	B. breve	B. cornutum	'B. deutium'	B. eriksonii	B. infantis ss. infantis	B. infantis ss. lactentis	B. infantis ss. liberorum	B. infantis - other	B. longum ss. longum
amygdalin	a	a	a^{w}	a	$-^{a}$	a^{w}	-	a	a	-	w^{a}	a^{w}	v_{w}	v
arabinose	v	a	a	a	-	$-^{w}$	-	a	-	-	$-^{w}$	-	-	a
cellobiose	a	a	a_{w}	a^{-}_{w}	$-^{w}$	a	-	a	a	$-^{w}$	v	a^{-}	v	v
erythritol	-	-	-	-	-	-	-	-	-	-	-	-	-	-
aesculin pH	a^{-}	v	a^{-}	v	-	a^{-}	-	a^{w}	$-^{a}$	$-^{w}$	-	v	v	-
aesculin hyd.	+	+	+	+	-	+	v	+	+	+	v	+	$+^{-}$	-
fructose	a	a	a	a	a	a	a	a	a	a	a	a	a	a
glucose	a_{a}	a	a	a_{a}	a	a	a	a	a	a	a_{w}	a_{-}	a	a
glycogen	-	a	v	-	-	a^{-}	-	a	a	-	-	a^{-}	-	v
inositol	-	-	-	-	-	-	-	-	-	-	a	a^{-}	v	-
lactose	a	a	a	a	a	a	-	a	a	a	a^{w}	a	a	a
maltose	a	a	a	a	v	a	w	a	a	a	a	a	a	a
mannitol	a	a	-	$-^{w}$	-	v	-	a	a	-	$-^{w}$	-	$-^{a}$	-
mannose	v	a	v_{a}	a^{w}	w^{-}_{w}	a	a	a	a_{-}	a	a	a^{-}_{w}	a	a
melezitose	a^{-}	a	-	v	-	v_{w}	-	a	a_{a}	-	-	-	-	a
melibiose	a	a	a	a	v	a_{-}	-	a	-	a	a	a	a	a
raffinose	a	a	a_{w}	a	-	a_{w}	-	a_{-}	a	a	a_{w}	a	a_{w}	a
rhamnose	-	v	-	-	-	-	w	w	-	-	-	-	-	-
ribose	a^{w}	a^{w}	a^{w}	v	$-^{w}$	a^{w}	w	a	a	a^{w}	v_{a}	a^{w}	v	a
salicin	a	a_{w}	a	v	-	a	-	a_{-}	a	v	-	a_{-}	v_{a}	-
sorbitol	a_{-}	-	a_{w}	-	-	v_{-}	-	w	-	-	-	a_{-}	-	-
starch pH	a_{-}	a	a	v	-	a	w	a	a^{-}	v_{+}	v	a_{-}	v_{+}	a
starch hyd.	+	+	v	v	$-_{a}$	v	+	+	v	-	-	v	-	+
sucrose	a_{w}	a	a_{-}	a	-	a	a	a	a	a	a	a_{-}	a	a
trehalose	-	a	a	v	$-_{-}$	v_{-}	w	a	a	$-_{w}$	v	a	v	-
xylose	a	a	a	a	w	w	-	a	a	-	a	a	v	a
gelatin dig.	$-^{w}$	-	-	-	-	$-^{w}$	w	-	-	$-^{w}$	-	$-^{w}$	-	-
milk	c	c	c	c	c	c	-	c	c	c	c	c	c	c
meat	-	-	-	-	-	-	-	-	-	-	-	-	-	-
indole	-	$-_{+}$	$-_{+}$	-	-	-	-	-	-	-	-	-	-	-
nitrate red.	-	-	-	-	-	-	-	-	-	-	-	-	-	-
catalase	-	-	-	-	-	-	-	-	-	-	-	-	-	-
human														
- clinical	.	+	.	+	.	+	.	.	+	.	.	.	+	+
- normal flora	I	IMU	IU	IU	IU	I	I	MU	M	IU	I	I	I	I

Table 11.9. *Lactobacillus*

	L. acidophilus	L. brevis	L. casei var. alactosus	L. casei var. casei	L. casei var. rhamnosus	L. catenaforme	L. cellobiosus	L. crispatus	L. delbrueckii	L. fermentum	L. leichmannii	L. minutus	L. plantarum	L. salivarius ss. salicinius	L. salivarius ss. salivarius
amygdalin	a^{-}_{w}	-	a	$-^{a}$	a^{w}_{w}	a^{-}	$-^{a}$	a_{w}	$-_{w}$	$-^{w}$	a^{w}_{w}	-	a^{w}_{-}	w^{-}	v
arabinose	-	a	-	-	-	-	a	-	-	v_{w}	-	-	a	-	-
cellobiose	a^{w}	$-_{w}$	a_{w}	a_{w}	a_{w}	a	a	w^{a}_{w}	$-_{w}$	-	a^{w}	-	a_{w}	-	$-_{w}$
erythritol	-	-	-	-	-	-	-	-	-	-	-	-	-	-	-
aesculin pH	a^{-}	-	w	a	w^{a}	$-^{a}$	-	$-^{w}$	-	-	a^{-}	-	a^{-}	w	-
aesculin hyd.	+	$-^{+}$	+	+	+	+	+	+	V	$-_{-}$	+	$-^{+}_{a}$	+	+	-
fructose	a	v	a	a	a	a	a	w	a	a^{-}	a	$-_{-}$	a	a	a
glucose	a_{a}	a	a	a	a_{w}	a	a	a_{-}	a	a	a_{w}	a^{-}	a_{w}	a	a
glycogen	-	-	$-_{w}$	$-_{w}$	$-_{w}$	a	-	w^{-}	-	-	-	-	$-_{w}$	-	-
inositol	-	-	-	-	-	-	-	-	-	-	-	-	-	-	-
lactose	a^{w}_{w}	-	$-_{-}$	a_{-}	a_{-}	a^{-}	$-^{w}$	w^{a}_{-}	$-_{-}$	a^{-}_{w}	$-^{a}$	-	a	a	a
maltose	a	a^{w}	a	a	a	v	a	w	a	a	v	-	a	a	a
aesculin hyd.	$+_{w}$	$-^{+}_{w}$	+	+	+	+	+	+	V	-	+	$-^{+}$	$+_{a}$	+	$-_{w}$
mannitol	-	-	a	a	a	-	-	-	-	-	-	$-_{a}$	-	a	a
mannose	a^{w}_{w}	-	a	a	a	a	a^{-}	w^{-}	a_{w}	V	a	-	a_{w}	a	a
melezitose	-	$-_{a}$	a	a	a_{w}	-	$-_{a}$	-	-	-	$-_{w}$	-	a_{w}	-	$-_{w}$
melibiose	v	-	-	-	$-_{w}$	-	-	-	-	v_{-}	$-_{w}$	$-_{w}$	a	a	a
raffinose	v	w^{-}	-	-	-	-	w	$-_{-}$	-	a^{-}	$-_{w}$	-	v	a	a_{-}
rhamnose	-	-	-	-	a	-	-	w	-	-	-	-	v	-	a
ribose	$-^{w}$	$-^{a}$	a_{-}	w^{a}	a_{-}	$-^{w}$	a_{w}	w^{a}	$-_{w}$	a^{-}	a^{-}_{w}	-	a^{w}	-	$-^{a}$
salicin	a^{-}	-	a	a_{a}	a_{w}	a	-	-	-	-	a	-	a_{a}	a	-
sorbitol	-	-	a	-	a_{w}	-	-	-	-	-	-	-	-	a	v
starch pH	a^{-}	-	-	-	-	a	-	a	-	-	a^{-}_{+}	-	v_{+}	-	-
starch hyd.	v	-	-	-	-	+	-	v_{-}	V	$-_{-}$	-	$-_{w}$	-	-	-
sucrose	a	$-^{a}$	a	w^{a}	a	a_{a}	a_{a}	a_{-}	a	a_{a}	a	-	a_{w}	a	a
trehalose	v	-	a	a_{-}	a_{-}	$-_{w}$	$-_{a}$	w_{-}	-	-	a	-	a_{a}	a_{-}	a_{-}
xylose	v	a^{-}	v	w	a	-	-	w	-	v	v	-	-	a	a
gelatin dig.	-	-	-	-	-	-	-	$-^{w}$	-	-	-	-	$-^{w}_{-}$	-	-
milk	c^{-}	-	$-^{c}$	c^{-}	c^{-}	$-^{c}$	-	c	-	c^{-}	$-^{c}$	-	c^{-}	c	c
meat	-	-		-	-	-	-	-	-	-	-	-	-	-	-
indole	-	-	$-_{+}$	$-_{+}$	$-_{+}$	-	-	-	-	$-_{+}$	-	-	$-_{+}$	-	=
nitrate red	-	$-_{+}$	-	-	-	-	-	-	-	-	-	-	-	-	-
catalase	-	-	-	-	-	-	-	-	-	-	-	-	-	-	=
human															
- clinical	.	.	+	.	.	+	.	+	+	+	+	+	+	.	+
- normal flora	IMU	IMU	IMU	IMU	IMU	IM	M	.	.	IMU	IU	.	IU	M	IM

Table 11.10. *Eubacterium*

	E. aerofaciens	E. alactolyticum	E. contortum	E. cyclindroides	E. lentum	E. limosum	E. moniliforme	E. nitritogenes	E. rectale	E. saburreum	E. tenue	E. tortuosum	E. ventriosum
amygdalin	-	$-^{a}$	-	-	-	-	-	$-^{w}$	v	-	-	w^{-}	$-^{a}$
arabinose	$-^{a}$	$-^{w}$	a^{w}	-	-	w^{-}	-	-	a^{w}	w	-	$-^{w}$	w^{-}
cellobiose	a^{-}	-	a^{-}	$-^{w}$	-	$-^{w}$	-	a^{w}	a^{w}	-	-	v	v
erythritol	-	$-^{w}$	-	-	-	a^{w}	-	-	$-^{w}$	w	-	-	$-^{w}$
aesculin pH	-	$-^{w}$	a^{-}	-	-	$-^{w}$	-	$-^{w}$	$-^{a}$	-	-	-	v
aesculin hyd.	$+^{-}$	-	+	+	-	+	-	+	+	+	-	+	+
fructose	a	a	a^{w}	a	-	a	a	a^{w}	a	w	$-^{w}$	w^{a}	a^{w}
glucose	a	a^{w}	a	a	-	a	a^{w}	a	a	a	w	a	a
glycogen	-	-	$-^{a}$	-	-	-	-	-	a^{-}	-	-	-	$-^{w}$
inositol	-	-	-	-	-	-	-	-	-	-	-	-	$-^{w}$
lactose	a^{-}	-	v	-	-	$-^{w}$	-	$-^{w}$	a^{w}	w	-	v	a^{-}
maltose	a^{w}	-	a	-	-	$-^{a}$	v	a^{w}	a^{w}	w	$-^{w}$	w^{a}	a
mannitol	-	a^{-}	-	-	-	a^{w}	-	-	$-^{a}$	-	-	-	-
mannose	a	$-^{w}$	v	a	-	$-^{a}$	a	a	v	w	-	-	a^{w}
melezitose	-	$-^{w}$	-	-	-	-	-	-	a^{-}	-	-	-	$-^{w}$
melibiose	$-^{w}$	-	a^{w}	-	-	-	-	-	a^{-}	-	-	w^{-}	v
raffinose	$-^{a}$	-	a^{w}	$-^{w}$	-	-	-	-	a^{w}	w	-	w^{-}	v
rhamnose	-	-	a^{w}	-	-	$-^{w}$	-	-	$-^{a}$	-	-	-	v
ribose	-	$-^{a}$	a^{w}	-	-	a^{w}	$-^{w}$	$-^{a}$	v	-	-	-	v
salicin	a^{w}	-	a	a^{-}	-	$-^{w}$	$-^{w}$	v	a^{-}	-	-	w^{-}	$-^{w}$
sorbitol	-	$-^{a}$	-	-	-	$-^{w}$	-	-	a^{-}	-	-	-	-
starch pH	-	-	v	-	-	$-^{w}$	$-^{w}$	$-^{w}$	a^{w}	w	$-^{w}$	-	w^{a}
starch hyd.	-	-	-	$+^{-}$	-	$-^{+}$	$-^{+}$	-	+	-	$-^{+}$	-	-
sucrose	a^{-}	-	a	a^{-}	-	$-^{w}$	-	$-^{w}$	a^{w}	a	-	a^{w}	v
trehalose	$-^{a}$	-	$-^{w}$	$-^{a}$	-	-	-	$-^{a}$	$-^{a}$	w	-	-	-
xylose	-	$-^{w}$	v	-	-	v	-	-	a	w	-	$-^{w}$	w^{a}
gelatin dig.	$-^{w}$	-	-	-	$-^{w}$	$-^{+}$	-	$-^{w}$	$-^{w}$	-	+	-	-
milk	a^{c-}	$-^{c}$	c	-	-	-	-	-	c^{-}	c	c^{d}_{d}	a^{-}	c^{-}
meat	-	-	-	-	-	-	-	-	-	-	-	-	-
indole	-	-	-	-	-	-	-	-	-	+	+	-	-
nitrate red.	-	-	-	-	$+^{-}$	-	v_{w}	+	-	-	-	v	-
catalase	-	-	-	-	-	-	-	-	-	-	-	-	-
human													
- clinical	+	+	+	+	+	+	+	+	.	+	+	.	.
- normal flora	I	.	I	I	I	I	.	.	I	M	.	I	I

Table 11.11. Antimicrobial susceptibility as a guide to the identification of some common non-sporing anaerobes (modified from Anaerobic Bacteriology Manual, Finegold *et al.*, see Appendix D)

	Anaerobic cocci Gram +	Anaerobic cocci Gram -	Gram +ve non-sporing rods (some)	*Bacteroides fragilis*	*B. melaninogenicus oralis ochraceus*	*B. corrodens*	*Fusobacterium mortiferum varium*	*Fusobacterium* other
Susceptibility to								
colistin 10 µg	R	S	R	R	V	S	S	S
erythromycin 60 µg				S	S	S	R	S^r
kanamycin 1000 µg				R	R	S	S	S
penicillin 2u	S	S	S^r	R^s	S	S	S^r	S
rifampicin 15 µg				S	S	S	R	S
vancomycin 5 µg	S	R	S^r	R	V	R	R	R

APPENDIX D

USEFUL REFERENCE BOOKS

Manual of Clinical Microbiology, Ed. Blair, J.E., Lennette, E.H. & Truant, J.P. (1970). American Society of Microbiology.

Anaerobe Laboratory Manual, Ed. Holdeman, L.V. & Moore, W.E.C. (1972). Available from and published by Anaerobe Laboratory, V.P.I. & S.U., Blacksburg, Virginia 24060, U.S.A.

The Pathogenic Anaerobe Bacteria, Smith, L.D.S. & Holdeman, L.V. (1968). Charles C. Thomas, U.S.A.

Isolation of Anaerobes, Ed. Shapton, D.A. & Board, R.G. Technical Publication No. 5 Society of Applied Bacteriology (1971), Academic Press, London.

Anaerobic Bacteriology Manual. Sutter, V.L., Attebery, H.R., Rosenblatt, J.E., Bricknell, K.S. & Finegold, S.M. (1972). Anaerobic Bacteriology Laboratory, The Department of Medicine, U.C.L.A. Medical Centre, Los Angeles, California, U.S.A.

Laboratory Methods in Anaerobic Bacteriology (1973). United States Government Printing Office, Washington D.C., U.S.A.

Anaerobic Infections, Finegold, S.M., Rosenblatt, J.E., Sutter, V.L. & Attebery, H.R. (1972). Upjohn Company-Kalamazoo, Michigan.

Anaerobic Bacteria: Role in Disease, Ed. Balows, A.,De Haan, R.M., Dowell, V.R. Jr. & Guze, L.B. (1974). Charles C. Thomas, Springfield, Illinois.

INDEX

Ecology of anaerobes, 59–64